JAGDISH MEHRA

THE QUANTUM PRINCIPLE: ITS INTERPRETATION AND EPISTEMOLOGY

D. REIDEL PUBLISHING COMPANY

DORDRECHT-HOLLAND / BOSTON-U.S.A.

Library of Congress Catalog Card Number 74-77965

ISBN-13:978-94-010-2236-1 e-ISBN-13:978-94-010-2234-7
DOI:10.1007/978-94-010-2234-7

Published by D. Reidel Publishing Company,
P.O. Box 17, Dordrecht, Holland

Sold and distributed in the U.S.A., Canada, and Mexico
by D. Reidel Publishing Company, Inc.
306 Dartmouth Street, Boston,
Mass. 02116, U.S.A.

Softcover reprint of the hardcover 1st edition 1974

Dedicated to

EUGENE P. WIGNER

with gratitude

CONTENTS

PREFACE

This study deals with the development of, and the current discussion about, the interpretation of quantum mechanics. The following topics are discussed: 1. The Copenhagen Interpretation; 2. Formal Problems of Quantum Mechanics; 3. Process of Measurement and the Equation of Motion; 4. Macroscopic Level of Description; 5. Search for Hidden Variables; 6. The Notion of 'Reality' and the Epistemology of Quantum Mechanics; and 7. Quantum Mechanics and the Explanation of Life.

The Bohr-Einstein dialogue on the validity of the quantum mechanical description of physical reality lasted over two decades. Since the early nineteen-fifties, Eugene Wigner has provided much of the point and counterpoint of the continuing discussion on the interpretation and epistemology of quantum mechanics. We have explored Wigner's views in some detail against the background of historical development and current debate.

Professor Eugene Wigner has sustained me over many years in my work on the conceptual development of modern physics by his kindness and encouragement. This study owes its existence to his direct inspiration, and to his suggestion to me in April 1971 that it would be of interest to write an account of the interpretation of quantum mechanics and the current discussion about it.

This study was completed in September 1972. Significant new developments have occurred since then in the discussion of questions related to the epistemology of quantum mechanics. Professor Ilya Prigogine has made decisive contributions to epistemology and the macroscopic level of description in quantum mechanics. (See I. Prigogine, Physique et Metaphysique, address at the bicentennial celebration of the Académie Royale de Belgique, 17 May 1973; Physics and Metaphysics, at the Ahron Katzir-Katchalsky Symposium, Berkeley, 1973.) Numerous other important considerations have been brought into focus by several distinguished authors (see the Section on 'Physical Description, Epistemology, and Philosophy' in J. Mehra (ed.), *The Physicist's Conception of Nature,* D. Reidel, Dordrecht/Boston, 1973).

I am very grateful to Professor Wigner for many conversations about the nature of physical reality and the interpretation and epistemology of quantum mechanics. I am also indebted to Professors Manfred Eigen, Joseph M. Jauch Léon Rosenfeld, and especially Professor Ilya Prigogine, for discussing with me their points of view about various aspects of the quantum mechanical description of nature. I am grateful to Dr. Helmut Rechenberg for giving me invaluable assistance in the preliminary organization of the literature on this subject.

I hope the present study would serve as a background of the continuing discussion on the interpretation and epistemology of the quantum principle.

JAGDISH MEHRA

CHAPTER 1

INTRODUCTION

The mathematical scheme of quantum mechanics (Q.M.) was established during the twelve months between June 1925 and 1926 by Heisenberg, Born and Jordan in Göttingen[1], by Dirac in Cambridge[2], and by Schrödinger in Zürich.[3] Important physical and mathematical contributions were made by de Broglie[4], Einstein[5], and Pauli[6], before and after the quantum mechanical scheme was established. The first step in the physical interpretation of the mathematical scheme was taken by Born who introduced the probability interpretation of the wave function.[7] The mathematical basis of the physical interpretation of quantum mechanics was developed further by Dirac and Jordan who independently worked out the transformation theory.[8]

With the confidence based on the results of transformation theory, Heisenberg explained the meaning of classical concepts in quantum mechanics and derived the uncertainty relations named after him.[9] Niels Bohr, one of the most important spokesmen of the philosophical *Zeitgeist* of the new quantum theory, arrived at his principle of complementarity, and sharpened its content in discussions with his collaborators, especially Heisenberg, in 1927. Bohr decided upon the principle of complementarity as the basis of his interpretation of Q.M.[10] From the work of Heisenberg and

Bohr there emerged the so-called Copenhagen interpretation which underlies the dominant orthodox philosophical view of Q.M.

Comprehensive attempts at formal and rigorous mathematical formulation and axiomatization of Q.M. were initiated quite early by Von Neumann[11], and they have been continued to this day by ardent adherents of mathematical purity in physical theory.[12]

Extensions of the Q.M. scheme followed immediately in the years after 1926/27. Dirac wrote the first paper on what is now called quantum electrodynamics.[13] Pauli, Klein, Jordan and Wigner developed and clarified the rules for field quantization.[14] In 1929 Heisenberg and Pauli laid the foundation of quantum field theory – the Lagrangian formalism, the pursuit of which led, after a struggle of many years, to the current quantum electrodynamics.[15] Application of Q.M. to the fields of atomic and nuclear physics brought a deeper understanding of the microscopic world, throwing new light on the structure of Q.M. itself, and the pioneering studies of Heisenberg, Bohr, Dirac, Wigner, Fermi, Yukawa and several others opened new vistas.[16] Bohr first introduced the ideas of Q.M. and certain aspects of its interpretation into more complex fields, such as biochemistry and biology[17], and in recent years Wigner has pursued and stimulated further investigations in this direction.[18]

The application and extension of Q.M. to new areas since the early 1950's have led to extensive discussions of its essential principles and their interpretation. Old debates have been renewed on account of the stimulus of fresh results, exhibiting a great variety among the points of views of the leading physicists. Among those who have developed and modified the orthodox Copenhagen interpretation of Q.M., Wigner's views have in particular provided point and counter-point of the new discussions. We shall analyze various responses, especially Wigner's, to the fundamental questions raised by the continuing research on the interpretation of Q.M., placing them, wherever possible, in the context of the historical background.

CHAPTER 2

THE COPENHAGEN INTERPRETATION

The Copenhagen interpretation (C.I.) of Q.M. grew from Heisenberg's 1927 paper in which he discussed the limitation on the simultaneous measurement of certain classical variables, such as position and momentum, in the new theory.[9] By devising thought experiments (*Gedankenexperimente*), he found that the accuracy with which certain (conjugate) pairs of quantities can be measured is limited. Bohr developed his unifying concept of complementarity on the basis of Heisenberg's examples. Beginning with his confrontations with Einstein at the Solvay conferences of 1927 and 1930, Bohr constantly sought to enrich the content of the complementarity principle, using it as a guide for the understanding of natural phenomena in full generality.[10,19] The role of the measuring apparatus in physical observations was discussed in detail, and it was formulated in strict mathematical terms by Von Neumann.[11] Around 1930, the basis of the orthodox interpretation of Q.M. consisted in the following statements:

(1) *Wave function:* The state of a physical system is described by a wave function ψ, considered as a vector in a Hilbert space.

(2) *Superposition:* If ψ_1 and ψ_2 represent two states of

a system, then a linear combination of the two also represents a state of the system.

(3) *Equation of Motion:* The wave function of a system obeys a continuous equation of motion, the Schrodinger-equation.

(4) *Measurement:* The process of measurement is described by the action of a corresponding Hermitian operator on the state. This action projects out an eigenfunction (of the Hermitian operator) and the related eigenvalue. The simultaneous measurement of two canonically conjugate (complementary) variables is subject to uncertainty limitations. In general, the process of observation brings about a discontinuous change of ψ, the state of the system, since it is not necessarily an eigenstate of the observable implying the "reduction of the wave packet".

Among the adherents of the orthodox Copenhagen interpretation of Q.M., including some of the most prominent names in physics, there exist no important or essential doubts concerning the validity of these statements. It must, however, be pointed out that the Copenhagen interpretation cannot be summarized as a set of logical axioms which are completely defined, because in many respects it is a *view* or a *school** rather than a *set of axioms*. This has several reasons. First, Niels Bohr, its most distinguished exponent, was

* Bohr never sought to create a 'school' around his philosophical ideas, in the traditional sense of the word. As for the *Copenhagen* interpretation of Q.M., Rosenfeld believes that C.I. "is not just one among many interpretations, it is the only one." (Discussion with Professor Rosenfeld.)

more of a Socratic philosopher rather than an Aristotelian or even a Platonist. He developed more a method of thinking which tells one how to attack and interpret physical problems, rather than a strict and narrow system (or some kind of foolproof program).[19] Heisenberg, who had developed the essential part of the theoretical scheme of Q.M. (which Bohr had not), accepted the complementarity interpretation of Bohr – but unlike Bohr, he never tried to formalize it.[20] Pauli, in his famous *Handbuch* article on the quantum theory, and in later writings, followed the Copenhagen interpretation as indicated in the above statements.[21] All three of them agreed that the wave function ψ describes a state of a microphysical system as completely as one can hope to describe it, that the wave function has to be interpreted statistically as shown by Born[7], and that one cannot hope to return to the classical description of nature.

A considerable clarification of the basis of this formalism has been accomplished since 1930. It was initiated by the important work of Birkhoff and Von Neumann[22], and elaborated and refined by a number of physicists and mathematicians.[23] In 1964 Piron[24] solved several of the problems left open by Birkhoff and Von Neumann. One of the most important of these problems was the justification of the complex Hilbert space for the representation of states and observables.

In the approach followed by Jauch and his collaborators, the fundamental concepts are the elementary propositions (yes - no experiments), and the physical structure of the quantal system is contained in the partial order relations which these propositions satisfy. Jauch and Piron have shown that these proposition systems always constitute a complete, ortho-complemented, atomic, weakly modular lattice.[25] Moreover, Piron has demonstrated that such a lattice has always a faithful representation in terms of the closed linear subspaces of a separable Hilbert space constructed over a field with a continuous involution.[24] Subsequent refinements of these results by Zabey, Eckmann, Piron and Gudder have established the physical reasons why this field is the complex number field.[26]

In another approach to Q.M. initiated by Jordan, Von Neumann, and Wigner[27], the fundamental objects are the observables which are represented by a certain algebra. In order to preserve the character of self-adjointness of the operators in Hilbert space which represent the elements of this algebra, it was considered necessary to postulate a symmetrical multiplication law which is not associative. Thus the symmetrical product of two self-adjoint operators,

$$A \circ B \equiv \frac{1}{2}(AB + BA) \tag{1}$$

is again self-adjoint and may, therefore, be considered as an observable.

Algebras with such a multiplication law have become the objects of considerable mathematical research, and their structure has been elucidated in detail. However, it has been found that they cannot provide a suitable basis for an axiomatization of quantum mechanics. This approach has therefore been replaced by another algebraic formulation in which the associative character of the multiplication is still maintained, but the observables are connected only with the self-adjoint elements of the algebras. This has led to the non-commutative C^*-algebra formulation of Q.M., with applications to field theory and statistical mechanics. The pursuit of these investigations, however, has not thrown new light so far on the fundamental problems connected with the axiomatic basis of Q.M.

During the several decades since 1930, other points of view have also emerged and various attempts have been made to modify the Copenhagen interpretation. The latter has always had its most vigorous champion in Rosenfeld, a long-time disciple and collaborator of Bohr.[28] In Rosenfeld's view of C.I., the concept of duality is as indispensable as Bohr's account of the measurement process given in 1927.[29] We shall see that not all physicists who consider themselves partisans of the Bohr-Heisenberg interpretation of Q.M. subscribe to the limitations set by Rosenfeld.

CHAPTER 3

FORMAL PROBLEMS OF THE Q.M. SCHEME

The Q.M. scheme described by the statements (1)–(4), which is accepted even by most of those who do not subscribe to the Bohr-Heisenberg interpretation, is not valid in all actual cases. Some parts of it need to be restated cautiously. These formal problems occur both in nonrelativistic quantum theory and in the free-field relativistic theory. To the latter, according to common belief, the current formulation of quantum mechanics should also apply. However, there exist problems as to what one might consider to be the accurate formulation and limitations of the Q.M. scheme.

3.1. Formal Problems in Nonrelativistic Quantum Theory

3.1.1. *Superselection Rules*

In 1952 Wick, Wightman and Wigner discovered that the superposition principle in quantum theory cannot be maintained without exceptions.[30] In Q.M., the usual assumption is that it is possible to carry out a 'complete' set of measurements, the result of which determines the state F completely but for the usual phase factor. If instead one assumes that the Hilbert space can be decomposed into certain orthogonal subspaces A, B, C, . . . , such that the components of F along A, B, C, . . . , namely F_a, F_b, F_c, . . ., have arbitrary phases with respect to each other, than no measurement can decide between

$$F_a + F_b + F_c + \dots \tag{2a}$$

and

$$e^{i\alpha}F_a + e^{i\beta}F_b + e^{i\gamma}F_c + \dots \tag{2b}$$

α, β, γ . . . , being arbitrary phases. By this fact the expectation values of any operator connecting at least two subspaces A and B, etc., will, in general, be completely undefined and such an operator will not correspond to a measurable quantity.

Now such 'superselection' rules which split the total Hilbert space of a system into orthogonal subspaces do, in fact, exist in nature. A trivial example is provided by momentum conservation. If all states with total linear momentum p are in a closed system, and there are no transitions between two states of different total momenta, the phase factors, α, β, . . . are unobservable. Further examples are provided by the conservation of the total angular momentum and the total electric charge. Wick, Wightman and Wigner showed that one can never distinguish between a state

$$f_A + f_B \tag{3a}$$

and

$$f_A - f_B \ , \tag{3b}$$

where f_A is a vector from a subspace with integral, and f_B a vector from a subspace with half-integral, total angular momentum.[31] They have recently returned to the case of the electric charge, because of some confusion which arose from recent papers.[32] In the latter, the argument was given that a system C, composed of two subsystems A and B, can evolve under a charge-conserving interaction in such a way that A changes from a state consistent with the superselection rule (say constant total charge) to another which is a superposition of different charges. Now Wick, Wightman and Wigner demonstrate that for this transition to happen, the system B has to be initially in a state which is a coherent superposition of states of different charges. Therefore, the superselection rules are not violated by these examples.

In the case of the charge superselection rule the proof proceeds as in the first example of "intrinsic parity" given by Wigner *et al.*[30] They construct the most general density matrix and then consider three processes: the union of two systems, the change of a system in time, and the separation of a composite system into two systems. The proof in the first two cases is straightforward: in the case of the union of two systems, one may consider the systems as statistically independent in the beginning and therefore take the direct product of the two density matrices as the density matrix of the composite system; the time change, on the other hand, is continuous and does not change the nature of the density matrix whether it agrees with the superselection rule or not. In the third process, however, where a composite system is decomposed into parts, one has to be careful because, on account of the possibility of an interaction between the systems, the density matrix of the composite system in general contains more information than the density matrices of the subsystems. The crucial point in this case is the fact that in talking about the density matrix in one system one has to decompose the Hilbert space of the total system; on introducing an orthogonal coordinate system one finds that the density matrix of either system is consistent with the superselection rule provided the density of the total system is also. Thus if we start with only 'permissible' states which are not superpositions of states with different charges, then we shall always end up with another 'permissible' state. The superselection rules, therefore, limit the superposition principle in an obvious and unambiguous way, such that one might call this limitation even trivial.

In the lattice-theoretic formulation of Q.M. none of these problems arise since, as Piron has pointed out, the lattice of propositions may decompose into a (discrete or continuous) union of irreducible lattices.[24] The states are then automatically mixtures of states in the irreducible sub-lattices, and nothing exists in this formalism which mathematically would correspond to the superposition from incoherent subspaces. Thus the lattice-theoretic formalism is better adapted to the actual physical situation since it avoids the use of irrelevant mathematical objects with no counterpart in reality.

3.1.2. *Are All Hermitian Operators Observable?*

The correspondence between observables and Hermitian operators in Hilbert space has not been entirely clarified. One usually assumes, more or less, a one-to-one connection between them.[11] In this connection, Von Neumann's statements can be summarized as follows:

(1) To every self-adjoint operator Q there corresponds an observable physical quantity.

(ii) The result of a measurement is always an eigenvalue q_ν of the operator Q.

(iii) The observation changes the state of a system ψ into an eigenstate ψ_ν of Q. The probability of the occurrence of the eigenvalue q_ν is given by

$$|(\phi_\nu, \psi)|^2 , \tag{4}$$

where (ϕ_ν, ψ) is the Hermitian scalar product of ϕ with ψ_ν.

The difficulty with the Von Neumann system (i)-(iii) is that it does not say how to carry out the measurement of the operator in practice. This question leads one into problems of the theory of measurement, and we shall treat it in detail later on. First, let us confine ourselves to a formal property of the measurement process which follows from the non-commutativity of the operators in Q.M. For this purpose it is sufficient to describe a measurement formally as

$$\phi\xi \to \sum_\nu (\psi_\nu, \phi)\psi_\nu \chi_\nu . \tag{5}$$

The initial state of the system is given by the vector ϕ, and that of the apparatus by ξ. The measurement yields the final states ψ_ν, indicated by the *macroscopically distinguishable* states χ_ν of the apparatus which correspond to the results q_ν. Now we may ask the question: For the operator Q, under what conditions are the states χ_ν of the apparatus macroscopically distinguishable?

Wigner was the first to mention explicitly that operators which do not commute with conserved quantized quantities like charge or angular momentum might not be measurable.[33] As a simple example, he chose a system in which the z-component of the spin is conserved, and asked for the measurement of the x-component of the spin. If, say, the two eigenfunctions of the z-component of the spin are ψ_0 and ψ_1, belonging to $-\frac{1}{2}\hbar$ and $+\frac{1}{2}\hbar$ respectively, then the eigenfunctions corresponding to the x-component are $\psi_0 + \psi_1$ and $\psi_0 - \psi_1$ respectively. Then, according to Equation (5), the measurement of the x-component can be written as:

$$\begin{aligned}
&(\psi_0 + \psi_1)\xi \rightarrow (\psi_0 - \psi_1)\chi\,, \\
&(\psi_0 - \psi_1)\xi \rightarrow (\psi_0 - \psi_1)\chi'\,, \qquad (6)\\
&\text{with} \quad (\chi, \chi') = 0\,.
\end{aligned}$$

By rearranging these transformations, one obtains

$$\psi_0\xi \rightarrow \left(\psi_0 \frac{\chi + \chi'}{\surd 2} + \psi_1 \frac{\chi - \chi'}{\surd 2}\right) \frac{1}{\surd 2}\,, \tag{6a}$$

and

$$\chi_1\xi \rightarrow \left(\psi_0 \frac{\chi - \chi'}{\surd 2} + \psi_1 \frac{\chi + \chi'}{\surd 2}\right) \frac{1}{\surd 2} \tag{6b}$$

for the changes of the eigenfunctions ψ_0 and ψ_1 under the measuring process described by Equation (6). Now the expectation value of the right-hand side of Equation (6a, b) is given by the sum of the expectation values for $(\chi + \chi')/\sqrt{2}$ and $(\chi - \chi')/\sqrt{2}$, and both values are equal. On the left-hand side, however, we expect a lower expectation value of the z-component of the spin for Equation (6a). Thus, either the z-component of the apparatus is infinitely large or one cannot measure the x-component of the momentum accurately. One is, of course, able to approximate this quantity in the sense that the measurement becomes the more accurate, the higher the content of the apparatus with respect to the conserved quantity. Wigner's proof has been generalized to include arbitrary operators by Araki and Yanase.[34]

3.2. Formal Problems in Relativistic Quantum Theory

We do not intend to address ourselves directly to the difficulties of defining a finite relativistic interaction, although they are in the background of our considerations. The formal questions have more to do with an adequate formulation of the process of measurement in relativistic quantum theory.

3.2.1. *Instantaneous Nature of Measurements*

In the theory of relativity, the coordinates do not have any meaning independent of the observation.[35] Now one may, if the distance is not too large (compared to the curvature of space), reduce the measurement of space-like distances to a measurement of time; that is, all one needs for the observation of coordinates is clocks.[36] For the microscopic theory, it is essential that one ought to be able to consider the clocks as being microscopic, meaning thereby that it is possible to consider the distinction between the macroscopic pointer and the clock as inessential, regarding only the light signals (with, say, one photon) as the important part of the clock. Q.M. imposes two conditions on a clock which is supposed to measure reasonable distances corresponding to time intervals up to $T = n\tau$, where τ is the accuracy with which one can measure time intervals. First, in the course of the measurement of the time interval T, the clock should pass through n orthogonal states at times τ, $2\tau, \ldots,$ $n\tau$, which tells us that the clock has a mass uncertainty of the order nh/T $(= h/\tau)$. Second, the clock should show the proper time even after having been read once, which means that the process of reading it should not deflect the clock too much from its original world line, a condition which is satisfied if the clock is heavy enough.[37]

Reading the clock is connected with the emission of a photon or light signal of duration τ, within the accuracy considered, which gives the clock an indeterminate momentum $\hbar/c\tau$. After a time interval T_2, the clock is at a distance of $\hbar T_2/Mc\tau$ from the point where it would have been if it had not been read. In order to measure the distance between two points 1 and 2 within the accuracy τ, the mass M of the clock has, at least, to be larger than

$$M > (\hbar c^2 \tau) \left(\frac{T}{\tau}\right)^{1/2} . \tag{7}$$

Equation (7) follows from a consideration of time measurements at both the points 1 and 2 respectively, and a comparison with the ideal measurement in which the clock does not obtain a recoil. Instead of the result of Equation (7), a rough estimate would give:

$$M > \left(\frac{\hbar}{c^2\tau}\right)\left(\frac{T}{\tau}\right) . \tag{8}$$

The difference between the values given by Equations (7) and (8) can be explained physically. The more stringent condition (8) requires that the wave packet of the center-of-mass of the clock be confined to a region of size $c\tau$ throughout the time interval T. The weaker condition (7) only ensures that the interval T is defined with sufficient accuracy within the wave packet which is spread out (in this case) to the size $c\tau(\sqrt{T/\tau})$, that is much larger than $c\tau$.

The next question is to verify the microscopic clock which measures with the greatest accuracy possible. One condition that must be satisfied is that the clock should have a very small extension l. Salecker and Wigner considered possible examples of clocks, such as an ensemble of atoms in an excited state which indicates the flow of time by the number of particles decayed.[36] The essential point in their discussion is how one may realize the 'reading', and this is most easily done by considering the motion of three particles (in a potential) with respect to each other. If the physical dimension of the clock is $2l$, then for the lower mass limit of the clock one obtains.

$$M > \frac{\hbar n^2 T}{l^2} = \frac{\hbar T^3}{l^2 \tau^2} = \left(\frac{\hbar}{c^2\tau}\right)\left(\frac{T}{\tau}\right) \times \frac{c^2 T^2}{l^2} , \tag{9}$$

where n is the number of stationary states which the clock has to go through in measuring the distance T. Since, with the definition of accuracy given in Equation (8), the spread-out of the clock is $l \approx cT/n$, the ratio between (8) and (9) becomes equal to n^2. Thus, the conditions under which a simple microscopic clock measures the time to an accuracy τ is even more severe than what follows from a general application of the uncertainty principle. However, Salecker and Wigner were able to construct a composite microscopic clock which consists essentially of an ensemble of oscillators with the frequencies $1/\tau$, $1/2\tau$, ..., $1/n\tau$. The restriction on mass then is essentially of the order given by Equation (8).

3.2.2. *Conflict Between Successive Measurements*

As in the preceding section, we shall also discuss here less the problems of physical interpretation of Q.M. than those of proper definitions. In quantum field theory (Q.F.T.) one starts by introducing field variables as kinematic and dynamic quantities.[38] Heisenberg attempted to introduce the limitation of the measurability of field quantities in a way similar to that of introducing kinematic quantities in the nonrelativistic theory.[39,40] However, a critical analysis by Landau and Peierls demonstrated that further restrictions in Q.F.T. were necessary.[41] This situation was clarified in the famous paper of Bohr and Rosenfeld.[42]

The essential point is that if one considers point charges in order to measure field quantities, certain obvious difficulties arise. The inconsistencies disappear if one uses test bodies whose linear dimensions are large compared to atomic dimensions, such that their charges can be considered spread out continuously over the whole body. In the formalism of quantum electrodynamics (Q.E.D.), the field quantities have to be considered not as point functions, but as functions of space-time areas which formally correspond to the average of the idealized (in the classical sense) local field components in these areas. By this process one arrives at some limitations concerning the unique description of the time sequence of measurements. It should, however, be emphasized again that the limitations which must be dealt with have nothing to do with the atomistic or elementary structure of matter, because we are discussing here the role of classical test bodies (of classical dimensions) which influence and are influenced by the fields according to the laws of classical field theory.

Bohr and Rosenfeld[42] concentrated on connected space-time areas whose volume V remains constant in a certain time interval T, and defined the field as

$$\mathscr{E}_{\mathrm{x}}^{(\mathrm{G})} = \frac{1}{VT} \int_T \mathrm{d}t \int_T \mathscr{E}_x \, \mathrm{d}V \,, \quad \text{etc.} \tag{10}$$

For the two regions I and II, one derives from the commutation rules for the electromagnetic field,

$$\begin{aligned} &\Delta\overline{\mathscr{E}}_x^{(\mathrm{I})} \, \Delta\overline{\mathscr{E}}_x^{(\mathrm{II})} \sim \hbar \, |\overline{\mathscr{A}}_{xx}^{(\mathrm{I,\,II})} - \overline{\mathscr{A}}_{xx}^{(\mathrm{II,\,I})}| \,, \ldots \\ &\Delta\overline{\mathscr{E}}_x^{\mathrm{I}} \, \Delta\overline{\mathscr{H}}_x^{\mathrm{II}} = 0 \,, \\ &\Delta\overline{\mathscr{E}}_x \Delta\overline{\mathscr{H}}_y^{(\mathrm{II})} \sim \hbar \, |\overline{\mathscr{B}}_{xy}^{(\mathrm{I,\,II})} - \overline{\mathscr{B}}_{xy}^{(\mathrm{II,\,I})}| \,, \ldots \end{aligned} \tag{11}$$

where

$$\bar{\mathscr{A}}_{xx}^{(\mathrm{I},\,\mathrm{II})} = -\frac{1}{V_\mathrm{I} V_\mathrm{II} T_\mathrm{I} T_\mathrm{II}} \int\limits_{T_\mathrm{I}} \mathrm{d}t_1 \int\limits_{T_\mathrm{II}} \mathrm{d}t_2 \int\limits_{V_\mathrm{I}} \mathrm{d}V_1 \int\limits_{V_\mathrm{II}} \mathrm{d}V_2 \times$$

$$\times \left(\frac{\partial^2}{\partial x_1\, \partial x_2} - \frac{\partial^2}{c^2\, \partial t_1\, \partial t_2}\right) \left[\frac{1}{r}\delta\left(t_2 - t_1 - \frac{r}{c}\right)\right],$$

and

$$\bar{\mathscr{B}}_{xy}^{(\mathrm{I},\,\mathrm{II})} = -\frac{1}{V_\mathrm{I} V_\mathrm{II} T_\mathrm{I} T_\mathrm{II}} \int\limits_{T_\mathrm{I}} \mathrm{d}t_1 \int\limits_{T_\mathrm{II}} \mathrm{d}t_2 \int\limits_{V_\mathrm{II}} \mathrm{d}V_1 \int \mathrm{d}V_2 \times$$

$$\times \frac{1}{c} \frac{\partial^2}{\partial t_1\, \partial t_2} \quad [\quad] .$$

From Equation (11), one recognizes that the average values of all field components in the same space-time area commute with each other and can be measured independently of each other with arbitrary accuracy. Thus the non-commutability of different field quantities (such as $\mathscr{E}$ and $\mathscr{H}$) in the usual formalism is due to the use of the concept of a point charge which does not allow for an accurate boundary of the measured area.[42]

A quantum field exhibits two features. First, it shows the so-called vacuum fluctuations. Second, the uncertainty products in Equation (11), or their square roots, define a critical field quantity $\mathscr{U}$ which is of the same order as the vacuum fluctuations in the case when the linear dimensions L are smaller than cT, and much larger if one considers large linear dimensions, as is usually the case. With a 'macroscopic' test body of charge density ρ one can measure the field strength in a volume V over the time interval T to the accuracy

$$\overline{\Delta\mathscr{E}}_x \sim \frac{\hbar}{\rho\Delta x \,.\, VT} \;, \tag{12}$$

if Δx is the accuracy of the position measurement of the test body, which, in principle, can be made arbitrarily small. However, in Equation (12) the 'field' effects of the test body have not been included. These correspond essentially to the vacuum fluctuations, and can be neglected in the case of large spatial extension V.

The electric field, say the x-component $\mathscr{E}_X$, can be measured with a test body of charge density P, which causes a momentum transfer in it, given by

$$P_x'' - P_x' = \rho\overline{\mathscr{E}}_x VT \;. \tag{13}$$

If the momentum of the test body is measured at time t', then we have to add on the right-hand side of Equation (13), a term $\overline{\mathscr{E}}'_x$ whose accuracy is of the order of

$$\Delta\overline{\mathscr{E}}_x^{(\mathrm{I},\,\mathrm{I})} = \rho_1 \Delta x V_1 T_1 \,|\overline{\mathscr{A}}_{xx}^{(\mathrm{I})}|\,. \tag{14}$$

For the minimal value of the product of the uncertainty the measurement of field components in two different space-time regions I and II, one now obtains

$$\Delta\overline{\mathscr{E}}_x^{\mathrm{I}}\,\Delta\overline{\mathscr{E}}_x^{\mathrm{II}} \sim \hbar\,|\overline{\mathscr{A}}_{xx}^{(\mathrm{I},\,\mathrm{II})} - \overline{\mathscr{A}}_{xx}^{(\mathrm{II},\,\mathrm{I})}\,|\,, \tag{15}$$

which is consistent with Equation (8).

From this discussion, the conclusion can be drawn that one can talk about the time sequence of observations only when the time intervals are fully separated, that is if (and only if) one of the quantities $r-c(t_1-t_2)$ or $r-c(t_2-t_1)$ is zero for all pairs of points of the area I and II. These considerations were further extended later on by Bohr and Rosenfeld, and Corinaldesi.[43] In this context, the critical question to ask is this: Does the answer given above follow completely the spirit of Q.M., a theory which connects successive measurements? The problem with Q.M. can be avoided by taking infinitesimal extension of the domains in time, but this can be done only at the expense of relativity. If an arbitrary volume in space-time is considered, it is difficult to see how one can define a field strength in its center without taking the interaction with the apparatus into account, because the measuring apparatus might very well have interacted with the field at earlier points of the space-time domain. One possibility, which might solve the problem, is to consider the light cone of a point in space time rather than a constant time cut.[44]

CHAPTER 4

THEORY OF MEASUREMENT AND THE EQUATION OF MOTION

We shall discuss the theory of measurement by first reviewing the more formal aspects of the problem, and shall return to its conceptual basis later on.

4.1. The Formal Scheme of Q.M.

The Q.M. scheme consists of two rather distinct parts: the equation of motion and the theory of observation. This distinction might most appropriately be emphasized by examining the changes which the statistical operator W, representing a state of a system, undergoes according to the equation of motion and the process of observation respectively.[45] If R is an observable and the ϕ_n form a complete set of eigenvectors, P_n represents the projection of the eigenvector ϕ_n and H the Hamiltonian of the total system, then the two changes in question are given by the equations

$$W \rightarrow W_t = e^{-iHt} W e^{iHt} \quad , \tag{16}$$

and

$$W \rightarrow W = \sum_{n} (\phi_n, W\phi_n) P_n \ . \tag{17}$$

Equation (16) represents a unitary transformation due to the equation of motion, whereas Equation (17), which gives the result of the measurement, does not. How can these two different changes be reconciled within one formalism? Is it possible to express microphysics in terms of either Equation (16) or Equation (17) only?

The orthodox (Copenhagen) answer to this question was most clearly stated by Von Neumann.[11] In his view, Q.M. applies everywhere with the Schrödinger Equation (16). The discontinuous process described by Equation (17), however, arises from the fact that one wants to say something about the changes of a system S, whereas the continuous Equation (16) applies only to the system S_1, which is made up of the system S and the measuring apparatus M. Now in order to determine the quantum state of S, one needs another observing system M_1, so as to register this state and study whether it obeys Equation (16). Thus one actually arrives at a system $S_2 = S+M+M_1$, and the game proceeds ad infinitum. Conventionally one is not really bothered by this situation, because with Von Neumann one can show that the result of the measurement on S does not depend on whatever system is finally treated with the "reduction of the wave packet" equation (17). Apparently, this situation seems to indicate that one needs an "ultimate observer" whom one can neither eliminate nor deal with the Schrödinger equation. This observer has often been related to an "unphysical" property, called "consciousness".[46] To many physicists, not necessarily of the Soviet (Marxist) school, such a conclusion means the denial of a completely objective world, and they are not inclined to accept such an interpretation. There are at least two obvious formal ways out of the dilemma. First, one may try to push the discontinuous Equation (17) into the background. Second, one may drop the continuous Equation (16) and work only with relations between the different measurements. Both paths can be followed, with greater or smaller philosophical consequences, and we shall deal only with the formal side of these attempts, with as few changes in the physics and its interpretation as possible.

4.2. Quantum Theory as the Theory of Observations

Formally the most simple attitude toward Q.M. is to regard it entirely as a theory relating successive measurements. Indeed there arise certain difficulties with respect to the relativistic theory, as we have noted earlier, but at least the nonrelativistic theory can be formulated unambiguously.[47] The measurement at time $t = 0$ is described by an operator Q. The content of the continuous equation of motion (16) can be extracted by requiring that the same measurement at time $t \neq 0$ is connected with $Q(0)$ by the equation

$$Q(t) = e^{-iHt/\hbar}\, Q(0)\, e^{+iHt/\hbar}\,. \tag{18}$$

Now one defines a projection operator P_j which leaves the outcome j of the measurement unchanged, but annihilates the other state vectors; $P_j(t)$ is related to $P_j(0)$ by an equation of the same structure as Equation (18).

With these definitions Q.M. can be reformulated in terms of the projection operators of successive measurements. Two subsequent measurements with the outcomes i and j

are connected by the equation

$$\text{Trace}\,(P_j' P_i)/\text{Trace}\,P_i = \text{Trace}\,(P_i P_j' P_i)/\text{Trace}\,P_i\,, \quad (19)$$

which gives the probability that the second measurement yields the outcome j, if the first one gave i. Of course, as mentioned above, this reformulation, although it puts the emphasis on the discontinuous change due to measurements, nevertheless includes the Schrödinger equation. But the concept of the state vector which should undergo the two different changes is simply eliminated here, and only the correlations between successive observations play a role. Margenau and Hill pursued a similar point of view.[48] They dealt, however, with two simultaneous measurements rather than a succession of these. Rather than radically suppressing the notion of a state, they use it, and finally tend to the opinion that the act of measurement converts the original state into a mixture, whereby the "projection postulate" of Von Neumann can be weakened. We shall deal with some aspects of such a view later on, but first let us turn to the claim that one can base Q.M. entirely on the continuous equation of motion.

4.3. Quantum Theory without Quantum Jumps

Schrödinger best advocated the dream of quite a few physicists who wanted to formulate atomic theory in such a way that the mysterious quantum jumps, which are obviously at variance with classical physics, would disappear.[49] Here we shall mention a much more modest claim, based upon the full acceptance of the quantum mechanical formalism, the only exception being the "reduction of the wave packet" theorem which necessarily seems to enter into the present formalism. Feyerabend took a step in this direction by attempting to reformulate Q.M. on the basis of the equation of motion and some qualitative statements concerning the measurement process.[50] He started with Equation (17), formulating it as,

$$P_\Phi \to \sum_n |\langle \Phi | \phi_n \rangle|^2 P_{\phi_n} = \Sigma P_{\phi_n} P_\phi P_{\phi_n} , \qquad (20)$$

where P_Φ denotes the statistical operator in the initial state Φ, and ϕ_n the eigenfunctions of the observable R. He called the process described by Equation (20) an 'incomplete' observation, because the reduction to a specific state ϕ_i has not yet been done since the result of the measurement has not been 'read' yet. He claimed, however, that this 'direct representation' of the observation fits neatly into a 'continuous' quantum theory.

To support his point of view, Feyerabend considered the measurement of a variable with two eigenstates ϕ_1 and ϕ_2 with, in general, unequal probability according to Equation (20), and required that *these states should lead to macroscopically different positions of the pointer of the apparatus.* He then divided the observation process into three steps:

Step 1. Interaction between the system S and the apparatus M: This is the only interesting step in classical mechanics and immediately leads to the result. In Q.M., it leads to the 'transition' into the mixed state given by Equation (20). But this state, call it P', is certainly different from the analogue in the 'classical' case, namely

$$P' \neq P_M = |\langle \Phi | \phi_i \rangle|^2 P_{\phi_1} P_A + |\langle \Phi | \phi_2 \rangle|^2 P_{\phi_2} P_B \ . \quad (21)$$

Here P_A and P_B correspond to the macroscopic state of M, referring to ϕ_1 and ϕ_2 respectively.

Step 2. Incomplete measurement and its reprèsentation: The formal transition is due to

$$P' \rightarrow P_M \ , \quad (22)$$

and it cannot be reduced to the (continuous) Q.M. equation of motion, since it happens in the macroscopic world. The transitions from the quantum description to it always involve approximations.

Step 3. Choice of an element of the mixture: This leads to a complete determination of the state of the system plus the apparatus, but it also involves approximations to the macroscopic world.

What has been gained by this procedure? Thus far, only a suggestion which involves the unknown transition from the microscopic to the macroscopic world. More specifically, Feyerabend illustrated his assumption that the states of the system ϕ_1 and ϕ_2 have to be classically distinguished. With this prerequisite he could write Step 1 as

$$|\phi_1\rangle \rightarrow |\phi_1\rangle' = |c_1 \phi_1\rangle e^{-ia_1} \ ,$$

$$|\phi_2\rangle \rightarrow |\phi_2\rangle' = |c_2 \phi_2\rangle e^{-ia_2} \ , \quad (23)$$

and

$$a_1 - a_2 \gg 2 .$$

Thus the interaction between the system and the apparatus, which causes the transition given by Equation (23), will also destroy the interference between ϕ_1 and ϕ_2, provided that all a_1, a_2 satisfying (23) are *equally probable.* Under these assumptions one obtains immediately from Step 1, P_M instead of P' to a very good approximation, and one does not have to assure the *jump* from P' to P_M, since the difference between them is very small to begin with. In Step 2, one introduces a macroscopic observer who will not see the difference between P' and P_M anyway (principle of the macroscopic observer).

Feyerabend's view is interesting, although two assumptions are involved which might partially imply what he wants to prove. For instance, one might ask: What do "macroscopically distinguishable" microscopic states mean? Since it is essentially by this notion that Feyerabend overcomes the hurdle of quantum jumps, a closer examination is necessary. One suspects that by applying the correspondence-like transformation given in Equation (23), the job is already done. This brings us to the second question which concerns the nature of the approximation to the classical description. By choosing it consistently with the correspondence principle, the Q.M. formalism enters immediately even if the conventional measurement theory is not mentioned explicitly. Finally, as Jauch has pointed out, the equations (17) or (20), which form the starting point of Feyerabend's considerations, already include part of the quantum discontinuity.[51]

4.4. Indeterminate State Vector of the Apparatus

It is possible to be less mysterious about the nature of the measuring apparatus by making the following assumption. Since the measuring device is macroscopic, consisting of many atoms having many degrees of freedom, one can relate to it an unspecified (microscopic) state. This possibility has been studied, but its consequences seem to contradict experience.[52] The important point is that we want to obtain finally a result in which the state of the apparatus is definite. As an example, let us choose a system S with discrete states $\sigma^{(1)}$, $\sigma^{(2)}$, ..., having eigenvalues $\lambda_1, \lambda_2, \ldots$. If we start with an initially sharp state $a \times \sigma^{(\nu)}$ of the system plus the measuring apparatus M, then the interaction will not change the microscopic state of the system $\sigma^{(\nu)}$, but it does change the state of the apparatus, or

$$a \times \sigma^{(\nu)} \xrightarrow[\text{(measurement)}]{} a^{(\nu)} \times \sigma^{(\nu)} \ . \tag{24}$$

Here the $a^{(\nu)}$, which refer to different positions of the pointer, are orthogonal and macroscopically distinguishable. No statistical element has entered the prescription so far. However, if we start from an initially unsharp state of the system, namely

$$a_1 \sigma^{(1)} + a_2 \sigma^{(2)} + \ldots ,$$

then the measurement changes the initial state to

$$a \times \left[\Sigma a_\nu \sigma^{(\nu)} \right] \rightarrow \Sigma a_\nu \left[a^{(\nu)} \times \sigma^{(\nu)} \right] . \tag{25}$$

Now on the r.h.s., a statistical correlation enters between

the state of the object $\sigma^{(\nu)}$ and that of the apparatus, and the simultaneous measurement of the two quantities (the state of S and the pointer position of M) always leads to a concordant result. Therefore, one measurement is unnecessary, because in the final state we never have

$$a^{(\nu)} \times \sigma^{(\mu)} \quad \text{with} \quad \mu \neq \nu \,, \tag{26}$$

and we look only at the pointer of the apparatus. Thus far the orthodox view is typically expressed by Equation (25).

An obvious modification of the orthodox view consists in attributing to the final state (after the measurement), not a state vector [r.h.s. of Equation (25)] but a so-called "mixture" of state vectors

$$a^{(\mu)} \times \sigma^{(\mu)} \tag{27}$$

with probabilities $|a_\mu|^2$. The state would then not be changed at all by the measurement if one would relate the expression (27) to the various possibilities of the world, and simply say that our final observation increases our knowledge.[53] That the Equation (25) contains more information than Equation (27) can be demonstrated with a modified Stern-Gerlach experiment, in which we do not measure the splitting of the beam but bring it once more to have interference. This effect would not follow from Equation (27). Of course, one might say that this is not the rule of the game because it is necessary that one should actually carry out the measurement before one inserts the second magnet; but then we have to introduce again the concept of the unfinished measurement and we have not gained anything. Consequently we might be led to the assumption that already the initial state of the apparatus is a mixture of $A^{(1)}, A^{(2)}, \ldots$, with probabilities $p_1, p_2, \ldots$, $A^{(\nu)}$ being orthogonal to $A^{(\mu)}$ for $\mu \neq \nu$. Starting from a sharp state of the system $\sigma^{(\nu)}$, with probability p_ρ, we obtain after the measurement,

$$\mathscr{A}^{(\rho)} \times \sigma^{(\nu)} \rightarrow \mathscr{A}^{(\rho\nu)} \times \sigma^{(\nu)} , \tag{28}$$

and starting from an unsharp measurement, with probability p_ρ, we get

$$\mathscr{A}^{(\rho)} \times \sum_\nu a_\nu \sigma^{(\nu)} \rightarrow \sum_\nu a_\nu \left[\mathscr{A}^{(\rho\nu)} \times \sigma^{(\nu)}\right] = \Phi^\rho , \tag{29}$$

with

$$(\mathscr{A}^{(\rho\nu)}, \mathscr{A}^{(\sigma\mu)}) = \delta_{\rho\sigma}\delta_{\mu\nu} .$$

That is, all states $\mathscr{A}^{(1\nu)}, \mathscr{A}^{(2\nu)}, \ldots$ describe the state $\sigma^{(\nu)}$.

Now the most general state, for which the originally measured quantity has a definite value λ_μ (this state being also coupled to some state of a definite pointer position, that is one of the states

$$\sum_\rho \chi_\rho^{(\mu k)'} \mathscr{A}^{(\rho\mu)} \quad \text{with} \quad \sum_k P_{\mu k}(\psi^{(\mu k)}) = |a\mu|^2),$$

is given by

$$\psi^{(\mu k)} = \sum_\rho \chi_\rho^{(\mu k)} \left[\mathscr{A}^{(\rho\mu)} \times \sigma^{(\mu)}\right] . \tag{30}$$

But the mixture of $\Phi^{(\rho)}$ cannot be taken for a state $\psi^{(\mu k)}$, according to Equation (30), if more than one $a \neq 0$. Thus it is not compatible with the equation of motion of Q.M. that a state of the system $S + M$, after the measurement, be a mixture of states, each with a definite position of the pointer. D'Espagnat has generalized this result to hold for more general measurements than those considered here.[52]

CHAPTER 5

TRANSITION TO THE MACROSCOPIC WORLD

It is considered beyond any doubt that one should be able to describe the macroscopic world with the equations of Q.M. The question is simply how to do so and where do the specific features of macroscopic matter come in. The theory of measurement has to take a standpoint about this question, and although one cannot claim to have arrived at a complete answer, there do exist certain indications in the right direction. The Bohr-Heisenberg approach (C.I.) presented the first scheme which explicitly showed the difference between the macroscopic and the microscopic descriptions, putting its finger on the gap which separates the two. The C.I. sought to bridge it by means of the principle of complementarity. It might, therefore, be appropriate to make some remarks about the connection between the microscopic and macroscopic descriptions in C.I. We shall then turn to the statistical view advocated by Born. In fact, one should be able to talk about the Q.M. of one microscopic particle even though in most cases one measures a many-particle effect. Various ways have been proposed to reconcile the present classical prescription with Q.M., but the conclusion which many people have reached is that one cannot reach the level of classical mechanics with the Q.M. formalism. The difficulties of explaining the measuring process without doing violence to the Q.M. equation seem to indicate that one would have to add something in order to obtain the classical description, even in the classical limit.

5.1. Complementarity

A favorite way of explaining complementarity starts with the consideration of the wave and particle properties of the light quantum. For at least one hundred years, it had been thought that radiation, in particular the visible light, was as perfect a wave process as one could imagine. In 1905 Einstein discovered that, under certain conditions, radiation behaves like spatially concentrated packets, similar to point particles. De Broglie generalized this hypothesis to include matter (which exhibits wave properties under suitable conditions). Since the empirical verification of his prediction one talks about the dualistic nature of matter and radiation. When Heisenberg discussed the meaning of the classical kinematical quantities as applied to atomic physics he found that their application was restricted.[9] From his *Gedanken* experiments, he found that the concept of a point particle describing an exact orbit made as little sense as that of the classical wave theory, if atomic dimensions were considered. Heisenberg decided that there were two possibilities of representing atomic processes: one could either describe the phenomena in terms of space and time and accept the uncertainty principle, or keep the causal relationships as expressed in the mathematical laws and accept the disadvantage that the physical description in space and time is impossible. Both descriptions, however, are related statistically to each other.[54]

Bohr expressed the same situation slightly differently. His approach was not rooted in the mathematical formalism, but he felt more strongly than the younger generation the dilemma of the wave-particle duality. He therefore developed his intuitive approach independently from Heisenberg.[55] Bohr recognized that the wave and particle aspects are contradictory, but also that both are necessary to give a correct and complete picture of nature. He did not merely accept a reinterpretation of the classical concepts, be it wave or particle and their relevant properties; rather he invented a logical instrument to describe the new situation, namely "complementarity". Bohr interpreted the uncertainty relations as the price one has to pay if one uses the classical description. In fact, no situation can arise in which both of the complementary aspects of the phenomenon show up simultaneously and rigorously. The contradiction arises from the fact that one has to measure the properties. The interaction between the object of observation and the measuring apparatus cannot be neglected, even in the limiting case, and one cannot therefore really sharply separate the properties of the atomic system from the effect of the measuring apparatus whose behaviour must be expressed in classical terms. Whereas Bohr applied the concept of complementarity to two different descriptions, Pauli did so to two quantities which do not commute, such as the position and momentum of a particle.[56] Bohr persistently sought to apply and extend his principle of complementarity in an intuitive way, for instance to defend the consistency of the new (Q.M.) description of Nature in his discussions with Einstein, but he never formulated it in terms of an equation.[57] The lack of a mathematical formulation of the complementarity principle makes it difficult to discover any serious flaw in Bohr's interpretation. A strong stand in its favor has been taken by some who came strongly under Bohr's philosophical and epistemological influence, but no more concrete formulation has been given by them.[58] For instance, Rosenfeld essentially discusses those problems of the interpretation of Q.M., the intuitive solution of which had already been proposed by Bohr, and it is difficult to argue against his position.[59] This, however, does not satisfy those who seek to find a less intricate way of expressing the situation in Q.M. This situation concerns less the foundation of the theory, than a closer description of the measurement process.

5.2. Statistical Description

A coherent way of talking about nature was proposed by Max Born. In an exchange of letters with Einstein he further developed the idea that all description of nature is finally statistical; one simply forgets this fact in classical physics due to an idealization which is really not permissible.[60] Born realized that the wave function has to be interpreted in a statistical manner, and this does not seem to be the case in any classical description. If one starts, however, with the fact that the natural laws represent our knowledge about objects rather than the objects themselves, a fact which Born considered to be the essence of quantum theory, one can always assume that this knowledge is unsharp and incomplete.

Born claimed, for instance, that determinism in classical theory is rather superficial, and does not stand strict examination.[61] For example, a particle moving between two walls with a small unsharpness in momentum will, after a finite time, be anywhere between the two walls, and one should describe this situation by a statistical distribution rather than by well defined coordinates. The accuracy with which one can measure the initial distribution in quantum theory is limited in a definite quantitative manner, and one draws the necessary conclusions from it.

In Born's description of nature, therefore, there is no contradiction between our equations (16) and (17), because there does not exist a deterministic description from the start. Bopp has tried to extend Born's ideas in order to build a statistical quantum theory of a particle.[62] The point here is not so much that the statistics which one has to apply agrees well in all cases with classical statistical mechanics; this is, in fact, not true, and both statistics join smoothly in the appropriate limits. It should be mentioned that the statistics of micro-particles and classical macro-particles indicated here is not identical with the (classical) statistical mechanics of many particle systems, a suggestion which underlies the belief that Q.M. offers an excellent example of the importance of collective phenomena.[63] In recent discussions about the foundations of Q.M., Aharanov *et al.*, for instance, have sought to make it clear that Q.M. applies to single particles, and all the phenomena which show up are due to single particles if interpreted correctly.[64]

5.3. Probability in Quantum Physics

As the above discussion shows, the emphasis in the problem of the interpretation of Q.M. has shifted significantly. In fact, the original considerations of Bohr and Heisenberg have been turned around. One no longer asks how microphysics should be described in terms of macrophysics, but rather how to describe macrophysics in terms of microphysics. Such a question had, in fact, already been raised by Einstein in 1916.[65] From his derivation of Planck's radiation formula using statistical laws, he obtained the classical (Jeans) limit of his theory. Einstein's work influenced Bohr to develop the correspondence principle.[66] Rather early in the development of the interpretation of Q.M., while dealing with the problem of the thermodynamical equilibrium of an n-particle system, Wigner also found the connection between the classical and quantum domains.[67] A unifying approach, containing both quantum and classical mechanics as special cases, has recently been suggested by Piron.[68] In this description the quantum theory with superselection rules is in a sense an intermediate case between classical and quantum physics.

It is clear beyond any doubt that the probabilistic interpretation of the quantum formalism is here to stay, quite apart from the question whether an ultimate deterministic theory of the microphysical phenomena will eventually be possible. Even if such a theory were possible, its detailed mathematical description would presumably be as useless as the detailed motion of individual molecules for thermal systems. It is, therefore, of interest to have a mathematically rigorous and complete description of the probabilistic part of quantum physics, as one has for classical physics. The development of such a quantal probability calculus has not yet been made although the need for it has been expressed several times.[69, 70]

The difficulty becomes explicit if one considers the fact that the observables in Q.M. are represented by non-commutative linear operators. This leads to a serious problem if one wants to define the joint probability for pairs of such operators. This problem was analyzed in some detail by Margenau and Wigner for position and momentum variables.[71,72] For this case, they concluded that no positive definite joint distribution function could be defined.

This situation suggests a reexamination of the classical probability calculus, with a view to adapting it to the actual physical situation which one encounters in micro-physical objects. That such an adaptation is possible and necessary has emerged most clearly from the lattice-theoretic formulation of Q.M.

In order to distinguish logically between the mathematical formalism and its application one should make a distinction, as Jauch has emphasized, between probability *calculus* and probability *theory*.[73] The former is a purely mathematical structure and need not have any particular interpretation. The latter is concerned with the application of this structure to specific problems of physics. The application may be justified by a critical analysis of the limits of validity, and this constitutes the essential part of a probability theory. The calculus as such, if it is well done, should be free from any ambiguity since it is a part of pure mathematics. It is also arbitrary, to some extent, and it is conceivable that several different calculi are possible each with its own characteristic field of application. The logical situation here is very similar to that encountered in geometry. Euclidean geometry is a mathematical theory which has a vast domain of applications in many parts of physics. But it becomes useless for navigating on the surface of the earth, for which it must be replaced by spherical geometry. Thus by changing the mathematical structure, the range of applicability can be extended to the quantitative description of new phenomena.

Until recently only one such calculus has been considered, that developed by Kolmogoroff. The basic concepts of the classical calculus are event space, probability measure, distribution functions and expectation values. They have been discussed extensively in the literature.[74] And until recently the applicability of this calculus has been assumed to be universal. However, this is not true, and the calculus which is needed for microphysics is not identical with Kolmogoroff's calculus.

There is an essential difference between classical and quantum physics, in the set of elementary events. Whereas in classical physics they form a Boolean lattice, and can therefore be identified with the subset of measurable sets, the lattice in quantum physics is not Boolean and cannot therefore be so identified. This fact has a decisive influence on the calculus of probability and we may call the new mathematical scheme quantal probability calculus.[75] Here a new measure arises which might be called a 'state' of the system, since the preparation of a state always produces a measure of this kind. The new concept of the measure does not modify the definition of random variables and distribution functions. But a crucial change is brought about by the fact that the distributive law of the proposition system is not valid any more. Hence there exist propositions which are not compatible, reflecting the non-commutativity of certain pairs of operators in Q.M. We do not wish to go into the details of all these modifications, but let us briefly mention a modification of a slightly different kind, that of using negative probabilities in Q.M.

The use of negative probability was first considered by Dirac with the goal of reducing the divergencies in quantum field theory.[76] This feature of the theory is intimately connected with the use of an indefinite metric in Hilbert space as Pauli showed in a review of Dirac's work.[77] An indefinite metric (indefinite form) in the space of states has been applied widely, especially in models of quantum field theory.[78] The question of the consistency of these theories is still widely discussed, but it seems that some progress has been achieved recently.[79]

We should emphasize, however, that the use of indefinite metric is ultimately linked with the relativistic aspects of

Q.F.T. In nonrelativistic Q.M. there is clearly no *need* of extending the notion of the metric. In fact, one can show under rather general conditions and assumptions, for instance those made in the axiomatic formulation of Jauch and his collaborators, that the Hilbert space with definite metric is inevitable.[80] Although several different but essentially equivalent representations of the lattice structure of the quantal proposition system are possible, in every class of equivalent realizations there exists exactly one canonical realization which uses a Hilbert space of positive metric.[81]

5.4. Macrophysical Description

The problem of the transition from quantum to classical description has often been discussed. One of its aspects is the passage from the probabilistic interpretation of the Schrödinger wave function to the classical deterministic evolution. A detailed analysis of the essential features of this problem was made by Moyal.[82] His conclusion was that the probabilistic features of quantum and classical statistical mechanics are different: classical statistical mechanics is a "crypto-deterministic" theory where each point in the event space evolves with a deterministic equation of motion; a similar interpretation of quantum kinematics is only possible with a probability distribution in phase space which cannot be assumed positive definite.[83]

The prescriptions for transition from the quantum to the classical description are well known in some respect, but it is still difficult to derive from these a formally satisfactory theory of measurement. An attempt in this direction was made by Ludwig.[84] The essential point in this derivation is the fact that classical and quantum physics are different theories, and can be related to each other only under specific circumstances.[85] For instance, in the Q.M. Hamiltonian function

$$H\left(\frac{\partial S}{\partial q} + \frac{\hbar}{i}\frac{\partial}{\partial q}, q\right), \tag{31}$$

one cannot simply let $\hbar$ tend to zero in order to obtain the behavior of a classical particle, because in general the operator $\partial/\partial q$ cannot be considered as small. By considering a free particle at high energies, it is possible to construct a suitable wave packet and obtain a particle which does not disintegrate, provided the uncertainty in energy is finite. The classical behavior, therefore, does not follow automatically from $\hbar \to 0$ or $E \to \infty$ or $m \to \infty$.

A simple notion of correspondence between the different domains of nature is not enough to furnish us with the essential features of the macroscopic description. Instead, one might examine the concept of a "macroscopic observable", and determine which among the quantum variables can be reinterpreted as macroscopic. Ludwig has discussed what certain prerequisites for the observables must be as a consequence of the ergodic theorem, that is, only those quantities play a macroscopic role that tend to a constant equilibrium value at $t \to \infty$ with reference to the internal variables and the center of mass, etc.[84]

According to Ludwig[86], the problem of the theory of measurement is the following: Let P be a projection operator for the variable A; its expectation value is given by

$$M(P) = \mathrm{Tr}\,(WP)$$

where

$$W_t = e^{-iHt} W_0 e^{iHt}, \quad \text{with} \quad \mathrm{Tr}\ W = 1 \tag{32}$$

gives the statistics, and H is the Hamiltonian operator of the system. Must one assume Equation (32) as an independent axiom in Q.M.? If so, how should one interpret its meaning?

Some other questions arise that must also be considered in detail. For instance: (i) Since W, which in a sense represents the quantum ensemble, is obtained from the experiment, one must ask how it should be defined. (ii) The variable A is measured with an apparatus which gives macroscopically fixed results. What is the significance of this fact for the structure of Q.M.? (iii) How does W change in the course of the measurement?

Crucial for all such questions is the relation of classical macrophysics to Q.M. By classical macrophysics Ludwig understands not necessarily a deterministic theory because, for instance, irreversible thermodynamics although classical is not deterministic. He considers classical physics and Q.M. as parts of a more general theory, non-existent so far, which would embrace both of them as approximate special cases. Because if Q.M. were to be considered as being always valid, then one would require a "supermacrocosmos" as the apparatus to test this assertion. The latter does not exist (on account of Von Neumann's final cut).[11]

What are the differences between the classical theory and

Q.M.? Certainly the axioms of Q.M., that every Hermitian operator represents an observable, and that any positive definite operator represents a possible ensemble breakdown in macrophysics. Now we know that even in the atomic world certain elementary principles, such as the construction of a product space say of two electrons, have to be restricted by the requirements of symmetry or anti-symmetry. In Ludwig's view, this process of increasing restrictions will continue to hold if one turns to more and more complex systems and finally arrives at a theory embracing Q.M., macrophysics and biology. Starting from the quantum systems, the same process also serves to determine the macroscopic variables. In the macroscopic case, there should exist a relatively much smaller number of degrees of freedom, as compared to those of the underlying quantum system. One requires that the time variation of the macroscopic variables be rather slow, and that they should always commute. The observables thus constructed are then related by correspondence to the classically known variables. In addition, the microscopic ensembles W change to $\overline{W}$, containing no interference terms, and satisfying the relation

$$M(A_{\mathrm{mac.}}) = \mathrm{Tr}(WA_{\mathrm{mac.}}) = \mathrm{Tr}(\overline{W}A_{\mathrm{mac.}}). \qquad (33)$$

A kind of smearing is finally carried out to reduce $\overline{W}$ to $\widetilde{W}$, a smooth function of the macroscopic variables $p_1 \ldots p_f$, which we denote by $\rho(p_1 \ldots p_f)$. One has therefore full knowledge of the measuring process once the transition of the ensemble W to the ensemble ρ is known.

The measurement of a quantum observable now consists

of the following steps: One takes a micro-ensemble W and constructs the variable $A_{\text{mac.}}$, and measures the latter macroscopically via the ensemble ρ. The whole problem is thus solved without any cut or introducing a subjective element. One must admit, however, that this procedure is rather tedious, and has to be repeated in all cases separately.

5.5. Measurement as the Increase of Information

Daneri, Loinger, and Prosperi have, during the last few years, tried to formulate a relation between statistical mechanics and the quantum theory of measurement, using essentially the idea that the collapse of the wave packet may result from an increase of information or an irreversible process involved in the act of observation.[87] They start from the assumption that Q.M. describes a microsystem only through its interaction with the macrosystem, and that statements concerning the macroscopic state of a large body have a meaning independently of its observation – that is, two persons who observe a macroscopic body or take a reading of the apparatus always "see the same thing". Accordingly, the problem is only to prove that the actual reading of an apparatus may be consistently interpreted as resulting in the increase of information.

In macrophysics the behavior of a body is described in terms of a limited number of (macroscopic) variables to which are ascribed well defined values at any time, and which evolve deterministically obeying certain differential equations. But a macroscopic body is made up of atoms to which the Q.M. description applies, including the concepts of *observation, probability* and *interference terms.* There are two problems in deriving macroscopic physics from Q.M.: (i) To construct Q.M. operators which describe macroscopic quantities, and (ii) to show that the dispersion of macroscopic quantities can be made negligible on the macroscopic scale, and that the (practically) well defined value of those quantities follows the correct laws of macroscopic physics.

The first question can be dealt with by the considerations of correspondence, but as Ludwig and his collaborators have shown, this problem is by no means trivial.[84,86] Daneri *et al.* are not bothered by this problem, and consider only the second question. Clearly, the macroscopic description of physical quantities is a closed one, and it is possible to calculate them at all times; the problem is to obtain this description with the help of a quantum ergodic theory of macro-systems. The macrostates are related to probability distributions over quantum microstates. The approach of Daneri *et al.*, more than that of Ludwig and collaborators, leads to the following solution: One should consider as results of the measurement only macroscopic quantities, obeying the macroscopic equations of motion. Underlying these, there exists a microscopic mechanics following microscopic laws which are eventually different from the macroscopic laws. The relation between the two is given by a master equation, similar to the quantum mechanical master equation.[88,89] The interference terms, which present a problem for the theory of measurements, are made to disappear by using the master equation and properly defining the term *macroscopic variable*. However, both of these assumptions go into this treatment of the measuring process, and one must remember that both are only *assumptions*, which are merely motivated rather than proved rigorously. Indeed, even in classical statistical mechanics, the existence of the ergodicity property remains to be proved.

Daneri, Loinger and Prosperi pursue their interesting considerations in detail. They have examined the specific master equation which should arise, examining it in the limit of macroscopic variables. A conclusion which arises from these considerations is that a macro-observer, for all practical purposes, may describe the behavior of the global system (consisting of a micro-object plus the macroscopic apparatus) by means of a given mixture at the end of the measuring process.[89] The interference terms are then practically absent, and the reduction of the state of the global system may therefore be interpreted consistently as merely the increase of information; and this occurs without the use of a second apparatus as in Von Neumann's treatment. One difficulty which is immediately apparent is what "practically absent" means, i.e. whether the measure of the situations described by "not practically absent" is negligible.[90] The proposals of Daneri *et al.* are still controversial.[91] A conceptual difficulty in their approach was emphasized by Wigner.[92] Since their formalism essentially includes the transition to the classical description, and implies the assumption that the apparatus finally is in a classical state, it shelves the problem rather than solves it. In addition, the terms "classical description" and "macroscopic theory" are not clearly defined, as can be seen from the examples of laser action and superconductivity current, which might be considered as *macroscopic quantum effects.*

5.6. Extension of Von Neumann's View

Von Neumann's formal theory has been further extended by Jauch and his collaborators since 1960.[12] Starting from the paradoxes stated in Von Neumann's description (see Equations 16 and 17), they try to find out the conditions under which one can arrive at a theory of measurement without passing on to the classical description, which seems to be required in other treatments of the problem. The result is that if a classical property already present in certain microsystems is given, then the discontinuous change of the wave vector can be obtained. Clearly, this classical property is a weaker condition than the requirement that the macrosystem (apparatus) behaves classically.

Jauch's starting point is the observation that a classical apparatus cannot separate all microscopic states, and the states which cannot be separated fall into equivalence classes. Moreover, the probability function which describes a physical (quantum) state is a more general object than the usual (classical) probability function. The latter is always defined on a Boolean lattice of classes of subsets, whereas the Q.M. probability does not even employ modular lattices. Now the Q.M. probability $P(E)$ of a projection E of a (separable) Hilbert space is of the form[93]

$$P(E) = \mathrm{Tr}\,(WE)\,,$$

with the linear operator W satisfying the conditions,

$$\begin{aligned}
&\text{(i)} \quad W^* = W\,, \\
&\text{(ii)} \quad W^2 \leqslant W\,, \qquad (34)\\
&\text{(iii)} \quad \mathrm{Tr}\ W = 1\,.
\end{aligned}$$

Moreover, W is Von Neumann's density operator having a discrete spectrum,

$$W = \sum_n \lambda_n P_n\,,$$

with

$$\begin{aligned}
&\text{(i)} \quad \lambda_n^* = \lambda_n\,, \\
&\text{(ii)} \quad 0 \leqslant \lambda_n \leqslant 1\,, \qquad (35)\\
&\text{(iii)} \quad \Sigma\lambda_n = 1\,,
\end{aligned}$$

where P_n is a projection operator of one-dimensional range. W is pure if $W^2 = W$. The statistical operator W for a pure state can naturally be related to the Schrödinger function ϕ, and

$$p(E) = (\phi, E\phi)\,. \qquad (36)$$

The interpretation of the classical and Q.M. states is now different, because quantum theory implies indeterminacy. In the case where the density operator satisfies

$$W = W^1 = \Sigma(\phi_n, W\phi_n)P_n = \sum_n p_n P_n\,, \qquad (37)$$

we have a classical description, and the projections of a statistical mixture can be measured without interfering with the state of the system. In a classical system we can always add conditions which reduce the dispersion of a state, but not so for the quantum systems – a fact which we shall discuss in the following section on hidden variables.

Now it is not clear whether all self-adjoint operators are observables. Some states might therefore not be completely determined, and one has to take the class of all states equivalent under all possible measurements as a *macrostate*. To these one ascribes *reality*. Every macrostate may contain many *microstates*. Again there exist pure macrostates and mixtures. Moreover, it is remarkable that the macrostates are invariant under observation. In a measuring apparatus, it is the classical aspect which is essential rather than its macroscopic nature, and the classical aspect is expressed by an Abelian set of variables. The analysis of actual measuring devices shows that there is always a lower limit to the size of the system representing the measuring apparatus in order to make it suitable as a measuring device. Although this lower limit has essentially no relation to the actual scale of human observers, it is usually of a size which would bring it into the category of "macroscopic" objects.

In order to render the measured event accessible to a conscious observer the measuring system must include a metastable amplifying device which can be triggered by events, such as the superheated liquid in a bubble chamber for instance. This device is necessary only to transform the measured event into a datum for an observer. This last step, as Jauch has noted, does not specifically have to do with Q.M., and is common to all classical physics as well. Jauch considers it unnecessary to pay any attention to the observer who transforms the event into a datum. In fact, from the measuring device M one needs consider only the classical but not necessarily macroscopic part m, and apply the foregoing considerations to the system S plus m as the total system. Jauch has discussed two kinds of measurements: in the first kind a repetition of the measurement yields the same result, while in the second it does not. In Q.M. one usually analyzes the first type of measurement.

Let S denote the system with a two-dimensional vector space spanned by the orthogonal vectors ϕ_+ and ϕ_-, while the state vectors of the system m span a three-dimensional vector space; ψ_0 describes the state of readiness of the device, ψ_+ the final state of m after it has measured ϕ_+, and ψ_- the state after it has measured ϕ_-. Then $\Phi_{+0} = \phi_+ \otimes \psi_0$ is the pure initial state vector before S and m have interacted, and after the measurement we have

$$\Phi_+ = U\Phi_{+0} = \phi_+ \otimes \psi_+ , \quad (U \text{ unitary}),$$

or (38)

$$\Phi_- = U\Phi_{-0} = \phi_- \otimes \psi_+ .$$

The general initial state now is actually Φ_0,

$$\Phi_0 = a_+\Phi_{+0} + a_-\Phi_{-0} , \qquad (|a_+|^2 + |a_-|^2 = 1) , \tag{39}$$

and the general final state, with Equation (38), is

$$\Phi = U\,\Phi_0 = a_+\phi_+ \otimes \psi_+ + a_-\phi_- \otimes \psi_- \,. \tag{40}$$

This pure state of the system $S + m$ becomes a mixed state if it is reduced to the measuring system m alone, because the reduction to m yields

$$W^m = |a_+|^2 Q_+ + |a_-|^2 Q_- \,, \tag{41}$$

where Q_+ and Q_- are the one-dimensional ranges containing ψ_+ and ψ_-. The reduction here is merely a mathematical consequence of the reduction of the pure state of Equation (40) to its component subspace. If the amplifying system is added to the measuring device, then one obtains the large classical system $S + M$, which is now a system and a measuring apparatus together with an amplifying device, and it automatically records a mixture rather than a pure state.

The result of these considerations is that the process of measurement can be explained entirely within the scope of Q.M. provided one is able to define the *classically distinguishable* states of the microscopic system, this being the only transition to the classical description remaining.[94]

5.7. Microphysics, Macrophysics and Dissipation

The problem of transition from the macroscopic level described by quantum dynamics to the microscopic level, at which the measuring apparatus operates, has been treated recently along a different path by Prigogine and his collaborators, following the concepts and techniques which they have developed in non-equilibrium thermodynamics and statistical mechanics.[95] Their principal idea is that the time evolution of systems may be split rigorously into independent *subdynamics*. One of these subdynamics, belonging to the projector $\widetilde{\Pi}$, contains all the information about the equilibrium and linear transport properties. Thus one might relate the macroscopic level of description of Q.M. to the reduced description in terms of the variable of the $\widetilde{\Pi}$-subspace. Such a description seems to be possible since an independent equation of evolution for the $\widetilde{\Pi}$-subdynamics can be obtained.

Thus one starts with a microscopic process which triggers a measuring device. After some time (of more or less macroscopic duration) one can study the evolution in the $\widetilde{\Pi}$-space and find that it behaves as one would expect from the "reduction" rule in Q.M.

The detailed analysis of these ideas consists in studying first the dynamics of the density operator in a superspace, given by a Liouville equation with the Liouville super-operator L. This Liouville operator can be derived from the Hamiltonian H as

$$L = H \times 1 - 1 \times H \,. \tag{42}$$

Here we have a direct product of operators in Hilbert space and its dual space. Splitting the Hamiltonian into

$$H = H_0 + V \,, \tag{43}$$

where the interaction V is responsible for "correlations", one might construct two subspaces of the superspace, characterized by the projection super-operators P_0 and P_c, with

$$\rho_0 = P_0\rho, \quad \text{and} \quad \rho_c = P_c\rho \ , \tag{44}$$

which are orthogonal.[95] The sub-densities ρ_0 and ρ_c are the average densities and the correlation amplitudes respectively, and they follow the coupled Liouville equations

$$i\dot{\rho}_0 = L_{00}\rho_0 + L_{0c}\rho_c \ ,$$

and (45)

$$i\dot{\rho}_0 = L_{c0}\rho_0 + L_{cc}\rho_c \ .$$

Extracting from Equations (45) the asymptotic forms $\tilde{\rho}_0(t)$ and $\tilde{\rho}_0(t)$ for large positive values of the time variable, one should be able to predict the further evolution on a macroscopic time scale. To do so, the Liouville super-operator is restricted to a class of systems which show normal asymptotic behavior, or an approach to a state of equilibrium. In these systems the effect of the correlation part L_{cc} should die out with time, leading to certain analyticity conditions for T_c,

$$T_c = \exp(-iL_{cc}t). \tag{46}$$

With these assumptions one derives the relation

$$\tilde{\rho}_c(t) = \int_0^\infty d\tau \, e^{-iL_{cc}\tau}(-iL_{c0})\,\tilde{\rho}_0(t-\tau) \tag{47}$$

between the asymptotic densities $\tilde{\rho}_c$ *and* ρ_0.

An asymptotic time evolution operator θ is introduced by writing $\tilde{\rho}_c(t)$ in the form

$$\tilde{\rho}_0(t) = e^{-i\theta t}\,\tilde{\rho}_0(0) \ , \tag{48}$$

and one can calculate θ by obtaining for it a functional equation from the first Liouville equation (45), and solving it by iteration.[96]

The total asymptotic density $\tilde{\rho} = \tilde{\rho}_0 + \tilde{\rho}_c$ is considered as an exact solution of the Liouville equation. It is obtained in the superspace by applying a projection operator P_a which retains only those correlation processes which have a long-time effect and manifest themselves on a macroscopic level. The operator P_a is unsymmetrical in time, and with its help it is possible to construct the operator $\tilde{\Pi}$ which projects on the subdynamics. However, $\tilde{\Pi}$ defines a time symmetric subspace of the superspace in which the asymptotic part of the time evolution, starting from any given situation, remains confined. Thus it clearly exhibits the features observed at the macroscopic level. The irregular fluctuations occurring on the atomic time scale are contained in the complementary subspace.

Since $\tilde{\Pi}$ is not factorizable, one cannot ascribe to it any state vector but only a density super-vector $\tilde{\rho}$. Thus the evolution in the subspace defined by $\tilde{\Pi}$ can be entirely described in terms of probabilities, and the "reduction of the wave packet" happens automatically. Physically speaking, the measuring apparatus is described in the $\tilde{\Pi}$-space (since it has *thermodynamic* properties); hence any phase relations in the initial state of the atomic system are wiped out.

The scheme sketched above, based on the considerations of Prigogine *et al.*, is highly suggestive in its possibilities. It provides a different and, in many respects, a complementary and wider approach than some of the other schemes we have discussed.[97]

CHAPTER 6

HIDDEN VARIABLES

Ever since the creation of Q.M., and even earlier, attempts have been made to invent a substructure which behaves causally and continuously, but nevertheless explains the quantum jumps by specific restrictions.[98] A historical development which certain recent researches find particularly appealing is the manner in which Boltzmann explained the properties of gases from the assumption of an atomistic structure.[99] The analogue which one has in mind is that although the classical gas theory satisfies thermodynamics (and the Second Law), the underlying kinetics of gas molecules is entirely symmetrical with respect to the exchange of the direction of time. The dynamics of the subsystems is therefore richer than their collective dynamics, and by analogy one wonders whether subnuclear dynamics could be continuous and possibly causal, even though the superimposed nuclear and atomic dynamics are not. Of course, every analogy is not convincing, especially since one cannot decide that the substructure of the atomic world should exhibit features which the macroworld, at least in the classical description shows. In any case, let us discuss the motivation for the speculation about theories employing hidden variables, before we turn to the specific nature of these speculations and the question whether they are compatible with the present knowledge.

6.1. Substratum as a Solution of Divergences in Q.F.T.

An interesting speculation, which first originated in a model used by Fürth, was taken up by Bohm and Vigier.[100] Fürth had shown that the classical Brownian motion provided an analogy to the uncertainty principle, the main formula being

$$\delta x = \sqrt{\overline{(\Delta x)^2}} = a\Delta t \; , \tag{49}$$

where $\overline{\Delta x^2}$ is the average fluctuation in time Δt, and a is a constant. The velocity of a particle (mass m) in Brownian motion is given by the relation $\delta v = \delta x/\Delta t$, hence

$$\delta x \, \delta p = ma \; , \tag{50}$$

where $p = mv =$ the momentum. In Equation (50), the product ma takes the place of Planck's constant. Although, Brownian motion is random and undetermined, the *real* motion is well determined by the molecular impacts. If it is assumed that the Q.M. level corresponds to the level of the theory of Brownian motion, whereas the atomic level in kinetic theory corresponds to a sub-quantum mechanical level (containing numerous qualitatively new and unknown *hidden variables*), then Planck's constant can be *derived* from the properties of motions at the sub-level. It would then represent some statistical feature related to the temperature of the sub-nuclear entities.

Now there are two kinds of Brownian motions, one occurring at normal conditions and the other connected with critical opalescence. Bohm thought that the latter phenomenon might be related to the facts concerning elementary particles. Empirical evidence suggests that the inner structure of particles has many degrees of freedom. If the *substratum* of one particle, which is supposed to oscillate rapidly with a high frequency, is subjected to an impact with high energy comparable to this frequency, then new particles might be created. The substratum, however, is now conceived to be such that the uncertainty principle does not hold anymore.

Such a breakdown of the statistical aspects in Brownian motion is in fact observed for very short time intervals Δt of Einstein's theory given by Equation (49). For small Δt the *uncertainty* limitation will not apply, and one obtains instead.

$$\delta p \, \delta x < ma \,. \tag{51}$$

In a similar way, one might expect that Q.M. also breaks down for short times and distances. The sub-particle and the sub-quantum mechanical level, therefore, should start approximately at the same point. The divergences which arise in the usual formulations of Q.F.T. can be viewed as a consequence of extrapolating Q.M. to *unjustifiably* short times. More specifically, the wave function represents, according to the Bohm-Vigier view, the substratum and the statistical state of motions in the sub-quantum; the phase in the sub-quantum level being given by some internal parameter (e.g. the angle) of the substratum. One can now derive the correct quantum conditions as de Broglie did in 1923. The relativistic invariance of this derivation suggests that there might not arise any difficulties in relativistic Q.F.T.

This speculation certainly has appealing features. One could perhaps relate the substratum to the quarks or partons, whose particle properties have not been substantiated in any case. One might also use the substratum as a kind of *prima materia*, out of which all elementary particles are made and which occur as the various excitations of this matter, as has been proposed by Bohm and Vigier. However, this speculation leads to serious questions on the non-relativistic level. Can one derive Q.M. from it? This question arises periodically and thus far no definitive answer has been given, notwithstanding various claims to the contrary.[101]

6.2. Von Neumann's Proof of the Nonexistence of Hidden Variables

Von Neumann, in his book, gave a proof that hidden variables are incompatible with the results of Q.M.[11], and during the following twenty years they did not recur in the discussion of Q.M. The question is whether one can introduce, in addition to the wave vector $|\Phi\rangle$ a set of coordinates (the hidden variables) in order to complete the *exact* (classical) description of a physical system, and rederive the results of Q.M. by averaging over the hidden variables. Von Neumann's assumptions are: (a) To each observable of a physical system there belongs a unique Hermitian operator in Hilbert space. (b) If the observable $\mathcal{R}$ is represented by the operator

R, then the observable $f(\mathscr{R})$ is represented by the operator $f(R)$. (c) The expectation value of a linear combination of operators is given by

$$\langle aR + bS + ...\rangle = a\langle R\rangle + b\langle S\rangle + ... \quad (52)$$

From these assumptions, Von Neumann derived the following results: (i) Given a statistical ensemble, then there exists a positive Hermitian operator U, such that for every observable R one obtains

$$\langle \mathscr{R}\rangle = \mathrm{Tr}(UR)\,. \quad (53)$$

(ii) There exist no dispersion free ensembles, that is those for which

$$\langle \mathscr{R}^2\rangle = \langle \mathscr{R}\rangle^2 \quad (54)$$

for all observables.

(iii) A statistical ensemble is homogeneous if and only if a vector exists with unit norm such that

$$U = P_{|\Phi\rangle} \quad (55)$$

where $P_{|\Phi\rangle}$ is the projection operator along $|\Phi\rangle$.

In the theories with hidden variables, one starts from the assumption that there exist dispersion free ensembles, and that the hidden variables cannot exist in homogeneous statistical ensembles. The results (ii) and (iii) are therefore incompatible with the existence of hidden variables.

Von Neumann's proof was slightly extended by Furry.[102] He studied the Einstein-Podolsky-Rosen (E.P.R.) paradox and noted that a *real* description in the spirit of E.P.R. was at variance with Q.M. rather than being an indication of its completeness.[103] Nothing further happened with this question until Bohm took it up in 1951.[104]

In 1963 Jauch and Piron improved Von Neumann's proof by strengthening his theorem in several directions.[105] Basing their argumentation entirely on lattice theory, they did not need any assumptions corresponding to Equation (53). Indeed a lattice does not even admit the possibility of the algebraic operations implied in Equation (52) insofar as the non-commuting observables are concerned. The crucial assumption in their proof was the following: If $p(a)$ is the real-valued functional representing a state of the system then

$$p(a) = p(b) = 1 \quad \text{implies} \quad p(a \cap b) = 1 \,. \tag{56}$$

This assumption has an obvious and plausible physical interpretation. It means that if the two propositions a and b are true in a given state, then the proposition "a and b" (represented by $a \cap b$) is true too.

There are two kinds of criticisms which can be directed against this assumption. Either one can doubt[106] the possibility of always constructing a proposition $a \cap b$ if a and b are given, a criticism which was effectively answered by Jauch and Piron in 1969[68], or one can question whether states associated with definite values of hidden variables satisfy the above requirement. This objection was made primarily by Bohm and Bub[106] and Bell.[107] In fact, Bell showed that such hidden variable theories are possible, and if they are of, what he calls *local character*, they must result in some slight deviations of the correlation-type experiments that are initiated by the E.P.R. *Gedankenexperiment.* Such deviations have been recently subjected to experimental tests in different types of experiments and they all have been found to violate the inequalities of Bell.[108] The most precise experiment of this kind is the one by Freedman and Clauser on the polarization correlation of two photons emitted by an atomic cascade of calcium.[108]

The no-hidden variable proof has been extended by Misra to Segal's axiomatics of Q.M. with C^*-algebras, and he arrived at a similar conclusion.[109] Finally, Gudder was able to rederive the Jauch-Piron result under still weaker conditions.[110] Thus the only hidden variable theories which might escape the Jauch-Piron-Gudder "disproof" are: (i) not deterministic, but having Q.M. as a "limit"; (ii) deterministic, but not being able to reproduce Q.M. completely.*

6.3. Bell's Local Hidden Variables

Bell's analysis assumes as little of the motivation of Q.M. as possible, but takes over all of its results. He asks whether

* J.S. Bell has pointed out to me that this conclusion may be incorrect as e.g. the de Broglie-Bohr theory (i) is deterministic, and (ii) reproduces Q.M. completely as regards the results of measurements. "The de Broglie-Bohm hidden variables are indeed not among the mathematical objects permitted by Von Neumann *et al.* That just shows, in my opinion, that those authors were insufficiently imaginative. Your conclusion would be correct if Q.M. were meant to stand not only for certain predictions about measurements, but also for some ruling on permissible mathematical form. Such a ruling would seem arbitrary to me, and most readers would not understand that it is implicit." [Private communication from J.S. Bell.]

Q.M. states can be regarded as ensembles of states specified further by variables, in such a way that given values of these variables together with the state vector determine precisely the results of individual measurements. For his theory, Bell assumed only that it be local; or, if at two distant space regions I and II there are two instruments, the results obtained from the first one cannot depend on the setup of the second. Assuming that both apparatuses make a binary decision or, $A(a) = \pm 1$, where A is the measurement and a an apparatus variable such as an angle, and $B(b) = \pm 1$. In the usual Q.M., one can only state that either the result (+) or (–) arises. Suppose that hidden variables exist, described by a continuous variable λ such that for certain values of λ the observation of A (or B) gives either +1, or –1, making the measurement entirely deterministic. With the probability density $\rho(\lambda)$ for the hidden variable λ, one finds the inequality

$$|P(a, b) - P(b, c)| \leqslant 2 - P(b', b) - P(b', c), \qquad (57)$$

where

$$P(a, b) \equiv \langle A(a, \lambda)B(b, \lambda)\rangle = \int d\lambda \rho(\lambda) A(a, \lambda) B(b, \lambda) \qquad (58)$$

are expectation values of the different variables. If one talks about, say a double Stern-Gerlach experiment, one apparatus being placed in the region I, and the other in region II, then choosing $b' = -b$ and $a \perp b$, and if θ is the angle between b and c, one finds from Equation (58)

$$|\sin\theta| \leqslant 1 - \cos\theta\,, \qquad (59)$$

an equation which, in general, is not satisfied. One must therefore conclude that there exists no local hidden-variable theory which gives the same result as Q.M. for a measurement of spin correlations of two particles.[111] Wigner has simplified Bell's proof, but we shall not go into the details.[112]

6.4. The Hidden-Variable Model of Bohm and Bub

At various times since 1951 Bohm has developed hidden-variable theories. In 1966 he presented a new model[106] which showed the following features: First, it reproduced the statistical predictions of Q.M., if one averaged over the hidden variables. Second, it automatically reduced the wave packet during the measurement in accordance with the influence of the hidden variables. However, for very short times after the measurement, it yielded a result different from Q.M.

The essential features of the model can be discussed by using a simple model of a quantum variable, taking the two values, say of a spin one-half system. One assumes that the general state vector ψ,

$$|\psi\rangle = \psi_1 |S_1\rangle + \psi_2 |S_2\rangle, \quad |\psi_1|^2 + |\psi_2|^2 = 1 , \qquad (60)$$

does not represent a complete description of the state, but has to be supplemented by an *additional* state vector $|\xi\rangle$,

$$|\xi\rangle = \xi_1 |S_1\rangle + \xi_2 |S_2\rangle ,$$

with (61)

$$|\xi_1|^2 + |\xi_2|^2 = 1 ,$$

where the components of $|\xi\rangle$ are the hidden variables, having a random behavior. The change in ξ_1 and ξ_2 is supposed to be slow, and nothing else happens during the process of measurement. The equations of motion for the components of $|\psi\rangle$ are now determined by the hidden variables, according to

$$\frac{d\psi_1}{dt} = \text{const.} \quad \frac{|\psi_1|^2}{|\xi_2|^2} - \frac{|\psi_2|^2}{|\xi_2|^2} \quad \psi_1 \, |\psi_2|^2 ,$$

and (62)

$$\frac{d\psi_2}{dt} = \text{const.} \quad \frac{|\psi_2|^2}{|\xi_2|^2} - \frac{|\psi_1|^2}{|\xi_1|^2} \quad \psi_2 \, |\psi_1|^2 .$$

From Equation (62) one finds, that the normalization of $|\psi\rangle$ remains constant. Now, if

$$\frac{|\psi_1|^2}{|\xi_1|^2} > \frac{|\psi_2|^2}{|\xi_2|^2} ,$$

one finally arrives at $\psi_2 = 0$ at the end of the measurement, and similarly at $\psi_1 = 0$ after the measurement if the inequality is inverted. One therefore obtains a reduction to the state $|S_2\rangle$ for certain values of the hidden variables ξ_1, ξ_2, and a reduction to $|S_2\rangle$ for the others. It can also be proved that Q.M. probabilities come out correctly. However, one can demonstrate that for times in which the hidden variables are not averaged, the predictions of the Bohm-Bub theory might contradict Q.M. This theory has been subjected to experimental tests and the result has been found to disagree with it.[113]

The conclusion from hidden-variable theories is not very encouraging, although one might wish to go around Bell's theorem by abandoning locality. Einstein, who throughout his life advocated the need of a *realistic* description of nature, pronounced a valid judgment on such theories when he wrote to Born in 1952: "Have you noticed that Bohm believes (as de Broglie did, by the way, 25 years ago) that he is able to interpret the quantum theory in deterministic terms? That way seems too cheap to me."[114]

CHAPTER 7

THE NOTION OF "REALITY" AND THE EPISTEMOLOGY OF Q.M.

The various attempts to interpret the process of measurement in Q.M. in terms of a unifying view, including the dualism of complementarity, have led to a new discussion about the deeper meaning of natural science. Before Planck's introduction of the quantum of action in physical theory, the belief in the principles of causality and continuity as the basis of natural science was deep and unshakable. The quantum jumps in atomic processes split some opinions, however. Even Einstein, who is often regarded (mistakenly, we believe) by the adherents of the C.I. of Q.M. as the very personification of classical determinism, thought that the very existence of atoms was at variance with the ultimate consequences of the classical view. Once a consistent description of the microscopic processes was known (or agreed upon), the question immediately arose as to how these processes and their description affect one's understanding of the relation between man and nature. The inventors of Q.M. first assumed a rather pragmatic point of view, almost akin to positivism. Niels Bohr sought to give the new science of Q.M. a more "humanistic" outlook from the very beginning, which can hardly be described in terms of positivism. Even if one does not follow Bohr in all details, one can seek to relate the description of the microworld to the human mind. Such subjectivism has been the subject of strong criticism by thinkers of the Marxist-Leninist school of materialism. Others have wondered about the emphasis on the description of a single object, and whether it can really be separated from the surroundings and the rest of the world. Both physical and philosophical questions therefore enter into the epistemological discussions of Q.M.

7.1. Description of Reality

For some time Einstein provided the essential counterpoint to the philosophical position adopted by the partisans of the new Q.M. He posed searching and critical questions to Bohr at the Solvay conferences of 1927 and 1930. By the time of the fifth Solvay conference in October 1927, Bohr had pondered very deeply on the questions of interpretation of the Q.M. scheme, and he could explain any *Gedankenexperiment* in the light of the Copenhagen interpretation. Bohr was able to meet all of Einstein's objections, and Q.M. and its interpretation were found to be internally consistent and free of contradictions. Later on, however, Einstein, Podolsky and Rosen (E.P.R.) raised another question: Can the Q.M. description of physical reality be considered complete?[103] In order to deal with this question, one must clarify the meanings of the words reality and complete. E.P.R. gave a necessary condition for their notion of completeness: Every element of physical reality must have a counterpart in physical theory. The reality condition is: A physical quantity may be considered to correspond to physical reality only if one is able to predict with certainty the value of the quantity without disturbing the system in any way.

In the wave description of a free particle one finds that when the momentum of the particle is known infinitely sharply, then the coordinate of position has no physical meaning. From this E.P.R. concluded that either, (1) the quantum mechanical description of reality given by the wave function is not complete or, (2) when the operators corresponding to two physical quantities do not commute, the latter cannot have simultaneous reality. In order to make their case, E.P.R. proposed the following example: Two systems I and II are permitted to interact during the time interval $t = 0$ and $t = T$. The states of the systems before $t = 0$ are known, hence one can calculate the state of the combined system I and II at any subsequent time with the help of the Schrödinger equation, in particular for $t > T$. One cannot, however, calculate the state of I or II after the interaction, but one has to measure it. Let A be such a measurement of the system I, with eigenstates $u_n(x_1)$ and eigenvalues a_n $(n = 1, \ldots)$, then one obtains

$$\psi = \sum_{n=1}^{\infty} \psi_n(x_2)u_n(x_1)\,. \tag{63}$$

The reduction of the wave packet in system I, say to $u_k(x_1)$, then leaves the second system, if separated after the interaction, in the state $\psi_k(x_2)$. The measurement of system I thus leaves only one term in the summation, Equation (63). Another measurement B on system I for $t > T$ may, however, leave system II in a state different from $\psi_k(x_2)$. Assuming A to be the measurement of the momentum, and B that of the position, one finds that a definite state cannot be assigned to system II. E.P.R. therefore conclude that the Q.M. description is not complete.

At various times Einstein returned to these considerations, and developed them further.[115] In particular he emphasized that objects should be arranged in space and time, and if they are situated at a given time in different parts of space, they should exist independently and not influence each other. Q.M. is in conflict with this requirement of reality. Is this requirement, however, really necessary to talk about reality?

Soon after the publication of the E.P.R. paper in 1935, Schrödinger was stimulated to present his views. The E.P.R. problem stemmed from the reduction of the wave packet, expressed by Equation (53), if a measurement is made on system I. In order to deal with this problem, Schrödinger formulated his famous "cat paradox".[116] He wrote:

It is possible to construct almost burlesque examples [to illustrate the situation]. Imagine a cat confined in a chamber containing the following 'torture device' which cannot be operated by the cat itself. A Geiger counter with a very tiny amount of radioactive substance is placed in the chamber, so that in an hour *perhaps* one of the atoms decays, with an equal probability that none decays. If an atom decays, then the counter reacts and, by a relay mechanism, triggers a hammer which releases a pellet of cyanide. On leaving the system alone for an hour one would say that the cat is alive *if* no atom has decayed in the meanwhile, while the first atomic decay would have poisoned it. The ψ-function of the whole system would express this situation by containing in itself the fact of the living and the dead cat mixed or smeared out in equal parts.

The typical situation in such cases is that the indeterminacy which is originally restricted to the atomic realm becomes a macroscopic sentient indeterminacy which can only be *decided* by direct observation. This prevents us from naively accepting such a 'washed out model' as a representation of reality. In itself, it does not contain anything unclear or contradictory, but there is a difference between a photograph which is unsharp or shaken during exposure and another one of clouds and layers of fog.[116]

Schrödinger had conceived a very colorful picture of the contradictions inherent in the description of reality. Many physicists, especially Einstein, were very happy about Schrödinger's example. Even as late as 1950, this question was of persistent concern to Einstein. In a letter to Schrödinger on 22 December of that year, he wrote:

You are the only contemporary physicist, besides Laue, who sees that one cannot get around the assumption of reality – if only one is honest. Most of them simply do not see what sort of risky game they are playing with reality – reality is something independent of what is experimentally established. They somehow believe that the quantum theory provides a description of reality, and even a *complete* description; this interpretation is, however, refuted, most elegantly by your system of radioactive atom + Geiger counter + amplifier + charge of gun powder + cat in a box, in which the ψ-function of the system contains the cat both alive and blown to bits. Is the state of the cat to be created only when a physicist investigates the situation at some definite time? Nobody really doubts that the presence or absence of the cat is something independent of the act of observation. But then the description by means of the ψ-function is certainly incomplete, and there must be a more complete description. If one wants to consider the quantum theory as final (in principle), then one must believe that a more complete description would be useless because there would be no laws for it. If that were so then physics could only claim the interest of shopkeepers and engineers; the whole thing would be a wretched bungle.[117]

7.2. The Complementary Description

It was quite natural that Bohr should take up Einstein's challenge and answer him immediately.[118] Bohr attacked the concept of physical reality, as stated by E.P.R., finding in it an essential ambiguity. He claimed that it is impossible to define the quantities, which E.P.R. had referred to, in an unambiguous way. Actually it depends upon the kind of measurement one performs or the manner in which one uses a certain apparatus, such as a slit, to determine the position and momentum. He therefore concluded that a physical quantity and its measurement are intimately related, and one is not independent of the other. Bohr stated that,

> In fact to measure the position of one of the particles can mean nothing else than to establish a correlation between its behavior and some instrument rigidly fixed to the support which defines the space frame of reference ... By allowing an essentially uncontrollable momentum to pass from the first particle into the support mentioned, however, we have by this procedure cut ourselves off from any future possibility of applying the law of conservation of momentum to the system consisting of the diaphragm ... .[118]

In fact, there *is* the question of an influence on the very conditions of defining the quantities.

Bohr's arguments against the E.P.R. definition of "reality", convinced most of his colleagues at that time. Serious doubts reappeared, however, when Bohm took up the challenge of constructing a deterministic quantum theory in 1951, and the arguments about the definition of reality were rejoined.

The philosophical discussion of the points of view of

Einstein-Podolsky-Rosen and Schrödinger, on the one hand, and Bohr and his followers, on the other, has recently been pursued further by Hübner.[119] He states these views more generally as follows:

> According to one principle, that of Einstein's, reality consists of substances which have properties, the relations they have to other substances notwithstanding. According to Bohr, however, reality intrinsically is a *relation* between substances, and measurement is a special case of such a relation. For Einstein a measurement *uncovers* a state in itself; for Bohr it constitutes a reality. For Einstein relations are defined by sunstances; for Bohr substances are defined by relations.[120]

Einstein and Bohr thus formulated two different philosophical principles, and they discussed the models or *Gedankenexperimente* within the range of application of their principles in order to find the consistency of their results.

As for Schrödinger's cat paradox, the argument involved is not necessarily cogent, because the state of the cat can be considered underlined in as much as it depends upon the atomic state of the radioactive substance. In contrast to this, however, the cat is actually dead or alive only in relation to certain medical (macroscopic) devices which can test its vital signs.[121] Einstein and Schrödinger are disproved only if one assumes Bohr's standpoint. The important philosophical or methodological question is to look for support of one standpoint or the other.

7.3. The 'Positivistic' Point of View or Two Kinds of Reality

When dealing with the reality of atoms and other microscopic objects one has to renounce some of the principles which apply to the ideal world conceived in terms of everyday experience. Einstein's theory of relativity set up the first restrictions, and concepts like rigid body and absolute time had to be given up. Atomic physics revised two other fundamental notions. The interference from the measuring device thus renders the general law of causality wrong; there exist in some way "uncontrollable" disturbances. One has also to give up the principle that all phenomena have to be explained as relations between objects existing in space and time. Clearly, there exists a body of exact mathematical laws, but they cannot be interpreted as expressing simple relationships between objects existing in space and time. There obtain two schemes in Q.M.[122]:

Either		*Or*
Phenomena described in terms of space and time but Uncertainty Principle present	Alternatives related statistically	Causal relationship expressed by mathematical laws, but physical description of phenomena in spacetime impossible

Let us recall that the main idea which led to Q.M. was to exclude from the theoretical scheme all those quantities which cannot be determined experimentally. Wigner has taken this aspect very seriously, and he has been led to consider Q.M. as a theory relating the results of successive measurements. He has stated that "the function of quantum mechanics is to give statistical correlations between the outcomes of successive observations. From a very positivistic point of view, such a reformulation is quite satisfactory: it refers solely to observation and establishes relations between them. From the point of view of everyday experience, in particular our belief in reality and the abstract characterizability of such reality, it appears to be very disturbing ... It appears that our theory denies the existence of absolute reality ... As to myself, I should admit that the old-fashioned concept of reality seems indeed old-fashioned and that I do not know how one could define operationally the reality of anything. My own thoughts, impressions and perceptions surely exist, the existence of all else is inferred only on the basis of these impressions and perceptions."[123]

In his article on 'Two Kinds of Reality', Wigner goes into further detail.[124] The first kind of reality is absolute; it is the existence of one's consciousness. The second kind of reality comprises the existence of everything else, in particular the concepts which are useful in describing the physical world, including the objects in it. One very often forgets about consciousness as a reality, because of too much concern with the external world. The nature of this reality is rather complex, "and the inadequacy of our appreciation of its properties may be one of the most potent barriers against the nature of universal realities at the present time".[125] As for things besides perceptions or the contents of one's consciousness, one might act as if they are real as long as they are useful. Thus even Q.M. concepts describe as much reality of the second kind as the classical ones.

Wigner's view seems to be both positivistic and subjectivistic, since he deals with things from the absolute standpoint of consciousness. He seems to seek to solidify the role of the subjective in science once and for all.

7.4. The Collectivistic-Materialistic View

The most vigorous opposition to "idealistic and decadent" Western philosophies has been expressed by Soviet thinkers, because in their official doctrine things have to exist objectively and atoms are like any other classical objects. The Soviet interpretation of Q.M. may well have started with the philosopher Alexandrov's book on 'The History of West European Philosophy'.[126] Alexandrov dealt specifically with the question of Q.M., and criticized the "bourgeois philosophy" which abuses the development of science for its own purposes. In his view, the Copenhagen interpretation (C.I.) has joined hands with Machian philosophy and fostered the development of subjectivistic concepts about the nature of Q.M.[127] One has to remember that Lenin himself, in contrast to the influential scientific opinion at the end of the 19th century and the beginning of the 20th century, had taken the side of Boltzmann in favor of the material existence of atoms against the school of energetics of Ostwald, Mach, and others.[128] Lenin found atomism in agreement with dialectical materialism, and he condemned Mach's positivism as representative of "bourgeois" and "decadent" philosophy. Later on, it was the duty of every good Marxist-Leninist scientist to treat Mach's ideas, and those of his intellectual successors, with doctrinal suspicion.

Blokhintsev's book on Q.M. gives a readily available account of the Soviet criticism of C.I.[129] According to this view, Bohr's principle of complementarity only emphasizes the possibility of macroscopic measurement of atomic objects, and not enough the fact that they are new. The principle of complementarity

> relates principally not to objective properties of the microuniverse, which result in the inapplicability of the methods of classical physics thereto, but to the possibility of an observer employing macroscopic quantities and concepts.
>
> This orientation of Bohr's complementarity principle leads to a twofold consequence. Firstly, Bohr, and his successors even more, have elevated this principle into a particular philosophical conception of complementarity, which leads them to deny causality and objectivity of microphenomena. On this basis Bohr speaks of the 'inadequacy of the customary viewpoint of natural philosophy for an account of physical phenomena of the type with which we are concerned in quantum mechanics', while Jordan declares outright the 'liquidation of materialism'.
>
> Secondly, the use of this principle of physics involves a subjective interpretation of the wave function and the concept of states in quantum mechanics. The wave function is regarded not as an objective characteristic of a quantum ensemble but as an expression of information gained by the observer as a result of measurement. The reality of any particular state of microsystems then becomes equivalent to information gained by the observer concerning such a system, i.e. becomes subjective instead of objective.[130]

Blokhintsev argues that C.I. is linked to positivism which considers neither the classical nor quantum physics as images of the objective world, but merely as mathematical constructs. The adherents of C.I., in his view, then analyze these constructs and reproach materialism because it would not give relations between subject and object. Blokhintsev states that true materialism acknowledges material relations between the cognizing subject and the object, which are expressed by the measuring apparatus. In addition, Q.M. studies the statistical laws of nature, and materialism asserts that the statistical law is an objective property of nature.

Single results are described only as they are related to the assembly, to the collective. The fact that a particle belongs to a pure or mixed ensemble, which possesses objective character and does not depend on the knowledge of the observer, is what determines the state of the particle. Viewed in this light, the E.P.R. paradox can be discussed as follows. The measurements of momentum p_1, and position x_1, in system I decompose the total wave function $\psi(x_1, x_2)$ into two different subensembles. The change in system II is therefore not due to the measurement, but due to the correlation of the two particles before the measurement. Hence E.P.R. were wrong in considering the microscopic particles as if they were classical. The seeming paradox, in Blokhintsev's opinion, arises only if one tries to understand the microscopic laws in terms of the laws of classical physics. But Lenin himself had stated that every scientific theorem on the structure of matter has only approximate validity, and new phenomena always appear which seem to contradict common sense, and this again only confirms dialectical materialism. Blokhintsev's comments perhaps do not represent the general opinion of the Soviet Russian physicists about the interpretation of Q.M. Fock, for instance, gave a faithful account of C.I., expressing his ideological bias in brief comments.[131] He rejected the so-called deterministic attempts of de Broglie, Bohm and Vigier as being too narrow. Fock himself arrived at a conclusion similar to the C.I., and stated that in the problem of the description of atomic objects "one cannot detach their behavior from the interaction with the measuring devices".[132] He made a distinction between the potentially possible concepts, including the wave function, and factual objects such as the measured quantities.

The entire complex of problems concerning reality, determinism, and theoretical constructs in quantum theory has been recently dramatized by Jauch in a Galilean dialogue on 'Are Quanta Real?'[133] In order to emphasize the similarity of the current ideological situation in Q.M. with the one prevailing in the 17th century, Jauch calls upon the three interlocutors Salviati, Sagredo, and Simplicio, created by Galileo to discuss the various aspects of the celebrated debate in his dialogue 'On the Two Major Systems of the World'.[134]

The central point in Jauch's dialogue (which, however, is never explicitly stated) is to show that the quest for reality cannot be solved in any consistent manner without taking at the same time into consideration the principle of complementarity. Reality, therefore, is not this or that object endowed with objectively given properties, nor is it the content of our consciousness or the idealized concepts which we use for imagining the inner coherence of elementary experiences. All these things are relative, as is shown by various striking examples treated in the dialogue.

Reality is found in the meaning, significance and relevance of a message, a perception or a concept within the context of a given situation. A variety of sense perceptions are received and registered by a process of filtering which separates those signals representing the noise from those containing the message. The important point is this: Different filters reveal different and complementary messages, all of which have equal claim to reality. Salviati considers this the essential content of Bohr's principle of complementarity without which Q.M. seems to give an incomplete description of reality.[61]

In the course of the dialogue on 'Are Quanta Real?' the three participants, starting from very specific examples in physics, extend their discussion to wider areas of scientific inquiry. Their debate reveals the universal significance of complementarity and its discovery in microphysics as one of the most important epistemological revolutions in the intellectual history of mankind.[61,134]

7.5. Can a System be Isolated?

The question whether one can isolate a system from the rest of the universe plays a great role in its description. In many cases, the concepts one uses apply only to such systems. A typical example is the Hamiltonian approach to dynamics, where in most cases one considers non-dissipative systems, and the "mechanical" description thus formulated is complementary to the thermodynamical description.[135] The thermodynamical description requires connection with a heat bath. i.e. to a system which is not described by the mechanics of the system in question. In fact, the very question of the increase of entropy arises from the fact that one uses two *incompatible* descriptions. In the case of Q.M., one could therefore argue that the programs pursued by Daneri *et al.*[86], and by Ludwig[87], in which the belief is expressed that only the state vector of the total world is an adequate description of any process, will not succeed. In fact, a cut in the sense of Von Neumann is possible only if the two systems can be isolated. But in principle, no system is completely isolated from the other, even though the coupling might be very weak.

Everett, Wheeler and Graham (E.W.G.) have proposed that the consideration of the state vector of the total world should be taken more seriously in order to get around the necessity of the collapse of the wave vector.[136] Under the assumption or postulate that the real world or any isolated part of it is represented by a vector, that its change takes place through dynamical equations, and that commutation relations are satisfied, E.W.G. need only another postulate to describe the results of Q.M. It is that the world must be sufficiently complicated, so that it is decomposable into systems and measuring devices. Here, one has obviously to prove not only that the statistical interpretation holds, but also how one should find the result to compare with the experiment.

The consequences of having a universal world state vector, with its myriads of components, would be vast indeed, but since we are part of the world we do not experience the world state vector itself. Still, one can play this game, which seems like gambling with the infinity, and find that the measurements turn out to be unique and the probabilities for a given result also come out all right. The difference between this and the usual formalism is that every outcome of the experiment, every possibility, is considered to be real. A difficult point which enters into this proof is that one has to introduce the probability of lower dimensional Hilbert spaces (subspaces). But the defenders of the E.W.G. system do not worry about this fact, because they believe that the statistics might go wrong only in those parts of the world which behave in such a way that no life or intelligence has developed there. At this point one might ask whether the observer (in the sense of Bohr) has in fact not been finally reintroduced. There are other problems with this view, for instance how one can finally observe anything at all, say in an atomic system. The arguments which are given in response are not fully clear.

The various views about reality that we have discussed have not led us to any specific new conclusion. Bohr's view, vague as it was, remains reasonable. To explain it in terms of Wigner's two types of reality would be interesting if it could help us in understanding something more than just ordinary Q.M. Wigner himself has attempted such an extension to one of the very old and fundamental problems, if not the most fundamental question man has ever asked: Can one explain the phenomenon of life? Or, more specifically, in our context: Can Q.M. explain life?

CHAPTER 8

QUANTUM MECHANICS AND THE EXPLANATION OF LIFE

Q.M. is usually regarded as a theory which applies to the micro-structure of matter. Many decisive processes occur in the atomic realm. The structure of molecules plays an important role in the functioning of living systems. It is therefore natural to ask whether Q.M. could provide a key to the understanding of life. In earlier times, there existed a common belief in the presence of a "life force" or some "vital" element in living substances which distinguishes them from dead or inanimate ones. However, serious scientific analysis of living organisms over a long period of time has not revealed the existence of a life force or vital element, this search being similar to that for the existence of a "heat substance" in matter in earlier times. In fact, an exclusion principle seems to operate in the very examination of life processes. In order to study accurately the composition and structure of an organism, say a bacterium, under an electron microscope, one has to kill it before one is able to do so. With the help of the physico-chemical apparatus one can analyze *physical* questions about the size, structure, and even the molecular composition of living systems, but this analysis eludes the question: What is life?

This was certainly the situation before the advent of Q.M.

From the foregoing remarks one might wonder whether all the knowledge gained from Q.M. would provide an essential change in the understanding of the fundamental question about the nature of life. It seems nevertheless that Q.M. has clarified certain concepts in biological phenomena, thanks to scientists like Delbrück and Elsasser who originally worked in atomic physics and went into molecular biology, participating in its new development. Among those who were active participants in the creation of Q.M., and who later on discussed biological questions from the epistemological standpoint, are Bohr, Wigner, and Jordan. We shall briefly examine how Q.M. considerations have affected the discussion of certain biological phenomena.

8.1. The Mind-Body Problem

In order to deal with Q.M. and the explanation of life, or some essential aspects of it, let us return to the process of measurement.

Wigner has extended his views about reality in order to understand the consequences arising from the measurement process in Q.M. and the appearance of "nonlinearity" in the description of microscopic objects.[137] He assumes that Q.M. has brought back into physics the recognition of mind as a primary reality.[138] First, so far as the results of observation of micro-objects are concerned, it does not matter who the observer is. Second, there exists a connection between the information from a past measurement and the behavior of the system in future, as we have pointed out earlier (see Section III).[139] The important point is that the result of an observation alters or modifies the wave function of a system, and this modification is

> in general unpredictable before the impression gained at the interaction [the measurement process] has entered our consciousness: it is the entering of an impression into our consciousness which alters the wave function because it modifies our appraisal of the probabilities for different impressions which we expect to receive in future. It is at this point that the consciousness enters the theory unavoidably and unalterably. If one speaks in terms of the wave function, its changes are coupled with the entering of impressions into our consciousness. If one formulates the laws of quantum mechanics in terms of probabilities of impression, these are *ipso facto* the primary concepts with which one deals.[140]

A subtle question arises if a "friend" observes a quantum system. In this case one has to describe, not the information available on the system as a wave function, but the total system, which is the *friend + object*. Now one can enter this system, which is a wave function even after the friend has observed, and ask him about his results. If he gives an answer 1, the wave function *friend + object* will separate and the wave function of the object will become ψ_1, if he gives the answer 2, the wave function will be ψ_2. Thus, whoever carried out the measurement, "the typical change in the wave function occurred only when some information entered *my* consciousness."[141]

Since thought processes become the primary concepts and the other kind of reality can, in principle, even be ignored, Wigner's point of view is radically opposed to any kind of materialism. It is obviously difficult to imagine how, with the present knowledge of physics and chemistry (and biology) the question as to what one's sensations are, and with what probability one out of several sensations is obtained, etc..., could be answered. Of course, there are connections between physical and chemical structures, between the content of a body and its sensations, and normally one assumes that the physico-chemical structure influences the sensations. According to Wigner, Q.M. tells us that the sensations might also influence the body. The reason is based upon the quantum theory of observation and measurement. There seems to exist no phenomenon in which a subject is influenced by another without exerting its effect upon the latter. Usually this effect is thought to be negligible, but this has never been confirmed.

The Q.M. argument runs as follows: Consider an object having two states, and an observer whom one asks for his observations. The subtle point now is that the observer whom we ask has already decided to tell us "yes" or "no" before we actually ask him. Thus it seems as though immediately after the measurement the system consisting of *object + friend* (the first observer) is already either in the state $\psi_1 \times \chi_1$ or $\psi_2 \times \chi_2$ (where ψ_i are the wave functions of the object and χ_i the wave functions of the observer), so that the state is not a linear combination of the wave functions. Hence there is a difference whether a physical apparatus makes the observation (which comes out finally in the superposition state) or a conscious being observes. A conscious being therefore plays "a different role in Q.M. than an inanimate measuring device".[142] As considered by Wigner, the breakdown of the superposition principle occurs here because ultimately a conscious mind gathers the observation. This is different from Ludwig's view, who claims that Q.M. does not remain valid even in the case of inanimate macrophycics.[86]

Wigner's view can be formulated in two theses. First, the so-called Q.M. measurement process is not complete until consciousness enters into the picture (and accomplishes the reduction of the wave packet).[143] Second, the consciousness plays a role different from an inanimate apparatus; it influences the physical world just as the physical world (structure) influences it (the consciousness). Can these two theses be verified experimentally? The difficulty in doing so arises because of the fact that consciousness is uniquely related to a specific physicochemical structure through which it alone receives impressions. Also, consciousness develops and becomes more informed (educated), as might be observed in a growing infant. Then there is the question whether consciousness modifies the usual laws of physics.

It is possible to formulate rationally the statement that it is the conscious being which completes the Q.M. measurement process. One may simply start from the assumption that the system *friend + object* cannot be described by a wave function after the interaction, but only by a mixture, i.e. the state is given either by the wave function $\psi_1 \times \chi_1$, with the probability $|\alpha|^2$, or the wave function $\psi_2 \times \chi_2$ with the probability $|\beta|^2$. As a result, Q.M. ceases to apply to a system with a conscious observer, because the final state requires a nonlinear description. It is, however, remarkable that there exists a continuous transition from the state described by the Q.M. wave function

$$\alpha(\psi_1 \times \chi_1) + \beta(\psi_2 \times \chi_2) \ , \tag{64}$$

and the non-Q.M. *state* described by the mixture

$$(\psi_1 \times \chi_2) \quad \text{with probability } |\alpha|^2 \ ,$$

and (65)

$$(\psi_2 \times \chi_2) \quad \text{with probability } |\beta|^2 \ .$$

The transition in this case, can be formulated by a 2×2 matrix

$$\begin{pmatrix} |\alpha|^2 & \alpha\beta^* \cos\delta \\ \alpha^*\beta \cos\delta & |\beta|^2 \end{pmatrix} \tag{66}$$

where the parameter δ has been introduced. If one takes $\delta = 0$, then the orthodox Q.M. situation is obtained, and the "statistical matrix", Equation (66), becomes singular. For $\delta = \frac{1}{2}\pi$, one obtains the mixture Equation (65). For intermediate δ, one has mixtures of two states given by Equations (64) and (65) with the probabilities

$$\frac{1}{2} \pm \left(\frac{1}{4} - |\alpha\beta| \sin^2\delta\right)^{1/2} \ . \tag{67}$$

The parameter δ looks like a new hidden variable, but it is quite different from those which we discussed in Chapter 6, because it relates together two different levels of description, the Q.M. and the complete or final statement about the observation. From this, as an immediate consequence, it would follow that Q.M. could not be applicable to the phenomena of life and consciousness, and a new explanation of life would have a profound effect on epistemology.[144]

8.2. Complementary Hierarchies

The Q.M. measurement process, the description and analysis of which, it would seem, is complete in the studies of Wigner, has led to the question of how to explain life with the new knowledge which has been gained about microscopic phenomena. Actually such an extension of Q.M. already started before Von Neumann's analysis[11] of the measurement process and, in a restricted sense, even before Heisenberg's fundamental paper[9] on the interpretation of Q.M. – in fact, even before the creation of Q.M. It was Bohr who thought very profoundly about the consequences of the quantum hypothesis, especially those which were emphasized by Einstein in the dualistic description of light.[145] At the time when Bohr developed his complementarity principle, he also had definite ideas as to how to extend it to a wider class of scientific phenomena.[10]

The reason why Bohr immediately thought of biological phenomena might have been two-fold.[146,147] First, his father, Christian Bohr, had talked to him about them. Second, Bohr had attended the lectures of the philosopher Høffding, a friend of his father, who was also a broad-minded thinker.[148] The elder Bohr, a physiologist, held the view that in biology one may have laws which, from a teleological point of view, have a certain meaning, but which, at least practically, one cannot trace back to physical and chemical origins. From Høffding, Bohr learned less about biological phenomena, than about the *scientific* basis and the *logic* of things. Høffding was not particularly good in technical analysis. In his lectures on the history of physics, he treated all theories at about the same level, laying emphasis on those problems which had not found solution in any of the existing systems. This attitude might have created in the young Bohr an open mind for *complementary* thinking.

Niels Bohr's thoughts on the relation between Q.M. and other (human) sciences are contained in his essays on 'Atomic Physics and Human Knowledge.[149] Almost all of these essays grew out of his lectures, and were written between 1932 and 1962. Already in his Faraday lecture in May, 1930, Bohr started to extend his ideas on the consequences of Q.M. to other sciences.[17] Addressing himself to the problem of life in a lecture on 'Light and Life' in 1932[150], Bohr expressed the belief that developments in atomic physics would have a new meaning for the understanding of living organisms within the general edifice of natural science.

Bohr discussed the construction and function of the eye and the reception of light. Since the absorption of even a single light quantum by the retina can lead to a visual impression, the sensitiveness of the eye may be said to have reached the limit set by the atomic character of the light processes, and the laws of Q.M. should be crucial for the understanding of biological phenomena. However, physical and biological descriptions are not directly compatible, if not mutually exclusive. There is a fundamental difference between biological and physical systems, the former being open and the latter essentially closed (or isolated, in the thermodynamic sense). While life has the typical aspects of self-preservation and self-generation, physical systems (in contrast) can be subjected to large numbers of subdivisions. Teleological argumentation plays an important role in biology as a complementarity feature, necessary to the description of life. Complementary points of view appear very clearly in two biological sciences, physiology and psycholoby. Bohr remarked:

> Without entering into metaphysical speculations, I may perhaps add that an analysis of the very concept of explanation would naturally begin and end with a renunciation as to explaining our own conscious activity.[151]

Bohr further developed this theme by recognizing that

> every experimental arrangement with which we could study the behavior of the atoms constituting an organism to the extent to which this can be done for single atoms in the fundamental experiments of atomic physics will include the possibility of maintaining the organism alive. The incessant exchange of matter which is inseparably connected with life will even imply the impossibility of regarding an organism as a well-defined system of material particles like the systems considered in any account of the ordinary physical and chemical properties of matter .... In this sense, the existence of life itself should be considered, both as regards its definition and observation, as a basic postulate of biology, not susceptible of further analysis, in the same way as the existence of the quantum of action, together with the ultimate atomicity of matter, forms the elementary basis of atomic physics.[152]

This, Bohr stressed, does not contradict any physical or chemical laws; and, in fact, the description of biological investigations is given almost entirely in terms of physics and chemistry.

Indeed, in Bohr's view, there is no contradiction between humanistic and natural sciences.[153] Complementary descriptions show up in both of them; as, for example, in anthropology, the two modes of behavior are characterized by *instinct* and *reason*. Disturbances through experiment and observation arise in humanistic studies also, such as those arising in the study of the culture of primitive people. In the context of the unity of all knowledge, Bohr referred to the example of free will which is often, albeit incorrectly, used in philosophical discussions about Q.M.[154] There is a contradiction between free will and total causality in the description of human actions. Q.M., Bohr said, is not able to account for the finalistic feature of organized purposefulness in living systems.[155,156]

The increase in knowledge about the structure of biological organisms and the processes occurring in them, which has accrued since 1930, inspired Bohr in his later years to return to the problem of the explanation of life.[157] Man's ability to construct machines which perform certain processes attributable to living organisms, such as self-repair, has strengthened the belief of some people that one would finally reach the goal of a *mechanistic* explanation of life. However, Bohr has pointed out that the complexity of living systems is far beyond that of any machine. The organisms

reveal properties and potentialities which are in striking contrast with those exhibited by so-called inanimate matter under simple experimental conditions. It is on this background that such notions as purposefulness and self-preservation, referring to the behavior of organisms as entities, have found fruitful application in biological research.[158]

Bohr proceeded to say:

> The basis for the complementary mode of description in biology is not connected with the problem of controlling the interaction between object and measuring tool, already taken into account in chemical kinetics, but with the practically inexhaustible complexity of the organism. This situation can hardly be regarded as being of temporary character, but would rather seem to be inherently connected with the way in which our whole framework has developed from serving the more primitive necessities of daily life to coping with the growth of knowledge gained by systematic research. Thus, as long as the word 'life' is retained for practical or epistemological reasons, the dual approach in biology will surely persist.[159]

This line of argument which, in some way, would appear to be a revocation of Bohr's earlier views, was carried on further in one of his last lectures.[160] He said:

> In reconsidering this conjecture [that life has to be taken as one complementary feature of organic structures] from our present standpoint, it must be kept in mind that the task of biology cannot be that of accounting for the fate of each of the innumerable atoms permanently or temporarily included in a living organism. In the study of regulatory biological mechanisms the situation is rather that no sharp distinction can be made between the detailed construction of these mechanisms and the function they fulfill in upholding the life of the whole organism.[161]

We have, in the foregoing, given a broad coverage to Bohr's ideas because he was, in the twentieth century, the first and most distinguished natural philosopher who attempted to tie together the different sciences: chemistry with physics, biological and humanistic sciences with physics and chemistry. Bohr is the founder of the modern unity of the sciences.

8.3. The Totality View of Life

The consequences of C.I. and of the *Kopenhagener Geist* of the late nineteen twenties, were pursued not only by Bohr, but also by younger men who carried out detailed investigations. Jordan, one of the principals in the creation of matrix mechanics, deserves special attention because he devoted himself, together with Zimmer, to the creation of the science of "quantum biology" in Germany.[162] Quantum biology arose from the study of processes in the smallest living systems, or in systems satisfying the essential functions of living organisms – these systems being the viruses and the genes.

Jordan considered the fact that single microscopic objects can cause biological changes, which means that the macroscopic explanation of life, the causal mechanism of Laplace, cannot be upheld.[162] On the basis of this result, Jordan concluded that the materialistic world-view is wrong.[163] In the realm of the smallest being which one might still call living, i.e. the virus, consisting of some millions of atoms, experimental observations concerning reproduction have been made. The latter research, partially due to Delbrück and his co-workers, has resulted in an understanding of the details of the reproduction process.[164] The starting point of the new reproduction theory was the assumption that a gene consists of one molecule[165], Q.M.

being necessary therefore to explain the most fundamental process of life. That is, the living organism is not, in its essential features, a Laplacian machine! Furthermore, from amongst the microscopic processes occurring in the molecule a steering mechanism transfers messages to the macroscopic dimensions (and parts) of the organism. Hence there is no contradiction between microscopic processes in the organism and their macroscopic consequences.

We must emphasize that, given Q.M., these considerations do not answer the central question about the explanation of life. When discussing complementarity, Bohr was not thinking of the details of how life processes proceed, as Jordan did. Biological studies of the 1930's had established the necessity of including Q.M. in the description of biological phenomena, and from this the totality view of living systems can be explained: that it is impossible to divide a living system into parts without killing it, i.e. without taking away the property which is characteristic of life. In macroscopic physics, divisibility is an integral part of phenomena.

8.4. Structure, Dissipation and Life

Having discussed the role of Q.M. in certain characteristics of living systems, let us consider some of the macroscopic aspects of the description of life. In the theoretical models of the functions of living organisms, one has to study those features which cannot be treated either by usual Q.M. or by the description of matter at equilibrium. It was pointed out long ago that biological organisms are open systems to which the mechanical equations do not apply. One might, however, seek to develop a statistical quantum mechanics, far from equilibrium, in order to discuss some of the features exhibited by living organisms. This problem has attracted much interest during the past several years, and we shall outline some of the main results.[166]

There exists an organization in thermodynamic equilibrium at low temperatures, as in the case of crystal structure, superconductivity, etc. We know that the organization of living systems is hardly of this nature, especially in that these systems interact with the external world and are therefore far from equilibrium. It is tempting to describe biological structures as open chemical systems operating beyond the stability of the thermodynamic branch. Two kinds of organization are necessary in order to keep the functions of living systems going on: organization in space *and* organization in time. Certain results, based upon simple nonlinear models, encourage the view that one might seek to obtain a physical description of these structures. The essential point is that there exist indeed new dynamical states of matter induced by a flow of free energy far from equilibrium. On a super-molecular level the states are governed by a new physical chemistry, while the laws referring to the molecular level remain essentially unchanged. Of course, coherent behavior on the super-molecular level corresponds to an amplification of specific molecular (Q.M.) properties under conditions far from equilibrium.

In all cases of these chemical stabilities and instabilities, a great deal of importance has to be attached to the interrelation between the system and its boundary conditions. The occurrence of a dissipative structure depends on the boundary conditions and vice versa. Dissipative structures are well defined in size and should, therefore, present a better model for biological structures than the usual systems on the thermodynamical branch. The possibilities of the fluctuations of both levels, i.e. the internal variables (describing the intramolecular state) and the macroscopic variables, may be present, and genetics might belong to the former. The models, even the successful ones, are very crude, requiring further specification, clarification and foundation, and one is far from the solution of specific problems. When this can be done, it promises to be an exciting task.

The question naturally arises as to what all this has to do with Q.M. and its ability to explain life. Moreover, what are the principles from which one can obtain a unified description of (inanimate and living) nature? This should be even more difficult than in the case of purely irreversible thermodynamics, which is related to the thermodynamic branch of the instability or fluctuation considerations of Glansdorff and Prigogine.[116] Here one must deal with a completely nonlinear statistical mechanics. One cannot solve any problem systematically, and must depend on models whose solution has been *accidentally* obtained. Even irreversibility cannot be derived from the mechanical or Q.M. equations alone, and essential additional assumptions have to be made to establish it. Similarly, in order to derive the new far-from-equilibrium dynamics that can deal with the description of life processes, additional postulates would be necessary.

We have already mentioned (see Section 5.7) the new attempt of Prigogine *et al.* to describe the macroscopic level of Q.M.[95] Prigogine's ideas on dynamical dissipativity, stability and evolution of structures in the framework of macroscopic level of Q.M. offer a new approach for the understanding of biological phenomena. The description of life may be obtained from these considerations with the help of the asymptotic subspace containing the uncorrelated density matrices. This would seem to contradict Q.M., where the correlations are thought to be important. Prigogine *et al.* believe that the functions of living beings can be described by the same subspace which can account for the measuring device. The living observer thus appears to function at the same level as the macroscopic measuring device.

In his conceptual scheme for the description of different levels of nature, Prigogine appeals to an intuitive hierarchy.[167] On top of the physical description of nature stands Q.M. We can obtain two main levels of description: The microscopic and the macroscopic. In order to obtain the latter, the essential effect of *dissipation* must be included.[95] At the macroscopic level, one has the possibility of applying thermodynamics, especially in equilibrium situations. Equilibrium may be reached from certain non-equilibrium states; other non-equilibrium situations are completely stable, and they correspond to dissipative structures. Thus, the latter deal with the primary conditions of life, with the possibility of reproduction provided by the genetic code, etc. Dissipativity leads to organization in space and time, which would disappear if new conditions allow the system to approach equilibrium. But if the system can maintain the ability to evolve through successive instabilities, then the preconditions for the establishment life will be fulfilled.[168]

Eigen has taken a slightly different position about the question of the explanation of life.[166] He does not attribute such a large role to spatial ordering within the system as Prigogine and his collaborators do. In his view, life has developed from molecular chaos to organization, but Eigen does not believe that

> macroscopic spatial structures are the key for understanding the first steps of biological self-organization. These steps may have occurred in a structureless 'soup', certainly involving functional macromolecular structures such as nucleic acids and proteins. The type of organization we need at the beginning is not so much organization in physical (i.e. geometrical) space. We need functional order among a tremendously complex variety of chemical compounds – possibly in a homogeneous phase. We need organization in a different 'space', which one may call *information* space.[169]

This order will also be based on the principle of Prigogine and Glansdorff to which Eigen assigns great significance, but in his view its utilization will require new parameters that will lead us beyond the realm of present thermodynamical theory.

8.5. Are We Machines?

Can physics explain life? Roughly speaking, there are two different views. One of these, that life can be explained entirely as a physico-chemical process, is quite popular among biologists. It seems that Prigogine and Eigen sympathize with this view. One must, of course, be careful and define exactly what is meant by physico-chemical laws. In the case of open or dissipative systems, this is by no means clear; certainly the physics of these systems does not coincide with the mechanical explanation of natural processes. The second view emphasizes the fact that something else has to be assumed in addition. Analysis based on Q.M. has shown that a fully *mechanistic* description of the measurement process is not possible. One has to go further and either assume a consciousness, as Bohr and Wigner have done, or a dissipative property as Prigogine *et al.* have proposed.

Wigner believes that mechanics, by means of physico-chemical processes, cannot explain the state of one's mind – whether one feels pleasure or pain.[170] A property which distinguishes living systems from machines is that the latter do not have anything like emotions or desires. The task is thus to supplement the *mechanical* description by suitable equations which denote the "state of the mind". It might be expected that these new equations would also modify the usual equations of Q.M.[171]

Although such a modified theory that would explain life is not yet available, we may look for some of its possible features. One of the most important is self-reproduction. Wigner has tried to estimate, on the basis of the present theory, the "probability of the existence of a self-reproducing unit".[172] The probability is zero, as can be proved by assuming that Q.M. applies. The proof goes like this:

Assuming that the "living state" is completely given in the Q.M. sense by a definite state vector with components ν_k, the system must contain at least one state (say, w) of the nutrient which permits the organism to multiply. Before "multiplication" the state vector of the *organism + nutrient* is

$$\Phi = \nu \times w , \tag{68}$$

($\times$ = direct product). After "multiplication" the state becomes

$$\psi = \nu \times \nu \times r, \quad \text{or} \quad \psi_{\chi\lambda\mu} = \nu_\chi \nu_\lambda r_\mu , \tag{69}$$

that is two organisms with the state vector ν plus a vector r describing the rest of the system. The indices χ, λ and μ describe the "parent" system, the "child" system, and the rest, respectively. The indices λ and μ should already be present in the nutrient system $w_{\lambda\mu}$, assuming finite dimensions N and R (for the indices χ, λ and μ respectively). Now the transition from Equation (68) to Equation (69) is described by

$$\nu_\chi \nu_\lambda r_\mu = \sum_{\chi'\lambda'\mu'} S_{\chi\lambda\mu;\chi'\lambda'\mu'} \nu_{\chi'} w_{\lambda'\mu'} , \tag{70}$$

i.e. a matrix S. The question now arises whether there exists any S which would do the job. For this purpose, let us take a given S and ask whether it is in general possible to find N numbers ν_k, R numbers r_μ, and NR numbers $w_{\lambda\mu}$, which satisfy Equation (70). From Equation (70), N^2R complex equations are deduced. Even if several identities exist among these, since there are only N (ν components) plus R (r components) plus NR (w components) free complex parameters, these are altogether much less than the number of equations. It is therefore extremely unlikely that the transition given by Equation (70) exists at all. Even if we assume that the living organism is not represented by one unique state ν, but a finite number of these, their number is still much smaller than the dimension of the state and we cannot obtain a finite probability for the transition described by Equation (70). One might, of course, make more realistic models of the reproduction process than Equation (70), but the outcome of zero probability remains unchanged.

Wigner chose the matrix S to be a random matrix. Eigen has criticized Wigner's calculation by pointing out that the assumption concerning the matrix S in Equation (70) might not hold.[173] In Eigen's view,

> Wigner's result r can actually lead only to the conclusion that any sophisticated state of matter which we now call 'living' cannot come about by random assembly – for a biologist certainly a trivial statement. The presence of instruction at the molecular level, which implies that the transformation matrix S must have a very specific form, may require an adaptation of statistical mechanics to special prerequisites of selective and evolutionary processes, but does *not* indicate 'that the present laws and concepts will have to undergo modifications before they can be applied to the problem of life' (E. Wigner).[174]

To this criticism, Wigner's rejoinder might well be: And how do you get instructions at the molecular level?

Wigner's ideas were slightly extended by Landsberg, using different conditions and assuming in particular that the number of degrees of freedom decrease rapidly in the presence of many living systems.[175] He obtained a probability different from zero, although very small. It has been pointed out that the reproduction property, as considered by Wigner and Landsberg, may not be peculiar to living systems; it also exists in inanimate systems, for instance in the neutron capture of atomic nuclei.[176] Wigner and Landsberg consider such examples as badly chosen, and believe that the "living state is a very uncommon state of matter of which it is composed, that is, that the same matter can be present in many, many other states which are not living".[177]

Wigner would propose that life and consciousness, together with the physical phenomena, should become the subject of common inquiry of a new unified science, and has wondered about the tendencies towards such unification.[178] The biology of lower organisms tends more and more towards the physico-chemical explanation, and it is indeed possible that in the near future these studies will be extended to organisms with consciousness. Q.M., on the other hand, in Wigner's view, deals with the connections between observations, i.e. in the contents of consciousness. Hence there is a need for developing a complete and unified description of the living organism, from biology and psychology, on the one hand, and from physics, on the other.

It appears very unlikely that the sensations experienced by sentient beings can ever be expressed in terms of the physico-chemical pictures. The fact that the laws of physics are formulated in terms of observation is strong evidence that these laws cease to be valid for the description of observations in which consciousness plays a decisive role. Wigner has therefore argued that the necessity for the modifications of the regularities obtained in the case of inanimate objects will become more and more apparent, leading ultimately to a harmonious synthesis of the two descriptions in one that is more complete.

APPENDIX

CLASSICAL AND QUANTUM PROBABILITY CALCULUS

1. Classical Probability Calculus

The basic concepts of the classical probability calculus are: the event space, the probability measure, the random variables, the distribution functions and the expectation values. We shall first give a brief review of these concepts and relations in the classical probability calculus and then point out the modifications which are necessary in the quantum calculus.

1.1. *The Event Space*

This is the basic object of any calculus of probability. In the classical calculus it is taken to be a *measure space*. It consists of a set Ω and a non-empty class $\mathscr{S}$ of subsets. This class is assumed to be a *field*. This means that it is assumed to be stable under the three operations of countable unions, intersections, and the formation of the complement. It is also assumed to contain the null set $\emptyset$ and the entire set Ω. In formulas,

$$S \in \mathscr{S} \quad \text{then} \quad S' \in \mathscr{S} \quad (S' \text{ complementary set to } S). \tag{A.1}$$

Furthermore, if

$$S_n \in \mathcal{S}, \quad (n = 1, 2, \ldots), \quad \text{then}$$

$$\bigcap_n S_n \in S \quad \text{and} \quad \bigcup_n S_n \in \mathcal{S}. \tag{A.2}$$

1.2. *The Probability Measure*

We assume that there is a positive-valued function μ which maps $\mathcal{S}$ onto the (closed) interval [0, 1],

$$\mu: \mathcal{S} \to [0, 1], \tag{A.3}$$

satisfying the following conditions

(a) $\mu(\emptyset) = 0, \quad \text{and } \mu(\Omega) = 1$ (A.4)

(b) The function should also be *σ-additive*, that is if $S_n \in \mathcal{S}$ is pairwise disjoint ($S_i \subset S'_k$ for $i \neq k$), then

$$\mu(\bigcup_n S_n) = \sum_n \mu(S_n). \tag{A.5}$$

The triplet $(\Omega, \mathcal{S}, \mu)$ is called *probability space.*

In the usual interpretation of this calculus, the sets $S \in \mathcal{S}$ represent possible events and the number $\mu(S)$ is the probability for the occurrence of this event under certain specified conditions.

1.3. *Random Variables*

Let χ: $\Omega \to \mathbb{R}$ be a real valued function: $\chi(\omega)$, $\omega \in \Omega$. For any set $\Delta \subset \mathbb{R}$ we denote by

$$\chi^{-1}(\Delta) = [\omega \mid \chi(\omega) \in \Delta] \tag{A.6}$$

the *inverse image* of the set under the function χ.

Further we denote by $\mathscr{B}(\mathbb{R})$ the class of Borel sets on the real line. A function χ is called *measurable-B*, or briefly *measurable*, if for every Borel set Δ the inverse image

$$\chi^{-1}(\Delta) \in \mathscr{S} \quad \text{for} \quad \Delta \in \mathscr{B}(\mathbb{R}) \ . \tag{A.7}$$

A *random variable* is a measurable function on Ω. We shall assume it to be real valued, but it is obvious that one can introduce equally well more general random variables.

In view of the generalization to the quantum probability calculus, it is worth pointing out here that the notion of random variable can be stripped to its abstract and essential content by noting that the function χ is not essential in the calculus, all that is really needed is the mapping which induces from the Borel sets on $\mathbb{R}$ into the field $\mathscr{S}$ of measurable subsets of Ω. It is useful to introduce a separate notation for this map and denote it by

$$\xi \colon \mathscr{B}(\mathbb{R}) \to \mathscr{S} ,$$

defined by

$$\xi(\Delta) = \chi^{-1}(\Delta) \quad \text{for all} \quad \Delta \in \mathscr{B}(\mathbb{R}) \ . \tag{A.8}$$

The map ξ satisfies the properties

(a) $\xi(\emptyset) = \emptyset \in \mathscr{S}$, and $\xi(\mathbb{R}) = \Omega$, (A.9)

(b) $\xi(\Delta_i) \perp \xi(\Delta_k)$ for $\Delta_i \perp \Delta_k$, and (A.10)

(c) $\xi(\bigcup_n \Delta_n) = \bigcup \xi(\Delta_n)$. (A.11)

Because of the properties (A.9) to (A.11) we denote such a

mapping as a σ-homomorphism of $\mathscr{B}(\mathbb{R})$ into S. Thus we may summarize: A random variable is a σ-homomorphism of the Borel sets into the measurable subsets $\mathscr{S}$.

1.4. *The Distribution Function*

Let ξ be a random variable, then the function

$$F_\xi(a) = \mu(S_a)$$

where (A.12)

$$S_a = \xi\,([-\infty, a])\,,$$

is called the *distribution function* of the random variable ξ in the probability space $(\Omega, \mathscr{S}, \mu)$. It has the following properties:

(a) $F_\xi(a)$ is a non-decreasing function of a, continuous from the right, with

$$F_\xi(a \to -\infty) \to 0\,, \quad \text{and} \quad F_\xi(a \to +\infty) \to 1\,. \quad \text{(A.13)}$$

(b) If the function $F_\xi(a)$ is absolutely continuous then its derivative satisfies

$$f_\xi(a) \equiv \frac{dF_\xi}{da} \geqslant 0\,,$$

and (A.14)

$$F_\xi(a) = \int_{-\infty}^{+\infty} f_\xi(x)dx\,.$$

The function $f_\xi(a)$ is then called the probability density. It exists only under the above special assumptions, while the distribution function $F_\xi(a)$ always exists.

1.5. *The Expectation Value*

Let ξ be a random variable and $F_\xi(a)$ its distribution function, then we define the integral (if it exists)

$$\langle \xi \rangle = \int_{-\infty}^{+\infty} a \, dF_\xi(a) \tag{A.15}$$

as the *expectation value* (or mean value) of the random varable ξ. If ξ is a random variable, then $(\xi - \langle \xi \rangle)^2$ is also a random variable and its expectation value is called the *variance* of ξ. The non-negative square root is the standard deviation $D(\xi)$ of ξ,

$$D^2(\xi) = \langle \xi^2 \rangle - \langle \xi \rangle^2 . \tag{A.16}$$

Of great importance in the application of the calculus is the notion of *independent* random variables. In order to explain this concept let us start with the preliminary notion of independent sets of $\mathscr{S}$. Two sets $A, B \in \mathscr{S}$ are said to be *independent* with respect to the probability measure μ, if

$$\mu(A \cap B) = \mu(A)\mu(B) . \tag{A.17}$$

This concept can be generalized to n sets $A_1, A_2, \ldots A_n \in \mathscr{S}$. They are independent if for every $i_1, i_2, \ldots i_n$, $(m \leqslant n)$,

$$\mu(A_{i_1} \cap A_{i_2} \cap \ldots \cap A_{i_m}) = \\ = \mu(A_{i_1})\mu(A_{i_2}) \ldots \mu(A_{i_m}) . \tag{A.18}$$

Two random variables are said to be *independent* with respect to the probability measure μ if for any pair $A, \mathscr{B} \in B(\mathbb{R})$ of Borel subsets on the real line

$$\mu(\xi(A) \cap \eta(B)) = \mu(\xi(A) \cdot \mu(\eta(B)). \qquad \text{(A.19)}$$

With this the essential concepts the classical probability calculus are explained.

2. Quantum Probability Calculus

Let us now proceed to the probability calculus in quantum physics.

2.1. *The Event Space*

In quantum physics the set of elementary events is a lattice of propositions, each of which is an equivalence class of yes-no experiments. In classical physics, this lattice has the property of being Boolean; therefore it can be identified with the subset $\mathcal{S}$ of measurable sets. In quantum physics, on the other hand, this lattice is non-Boolean and cannot be so identified. This fact has a decisive influence on the calculus of probability. We may call it quantal probability calculus to distinguish it from the classical calculus.

The quantal proposition system will be denoted by $\mathcal{L}$. It is a complete ortho-complemented lattice isomorphic to the closed linear subspaces of some Hilbert spaces.

2.2. *The Probability Measure*

A quantal probability measure μ is a function $\mu: \mathcal{L} \rightarrow [0, 1]$, satisfying the conditions

$$\text{(a)} \quad \mu(\emptyset) = 0, \quad \text{and} \quad \mu(\mathrm{I}) = 1\,, \qquad \text{(A.20)}$$

(b) $\mu(\bigcup_i a_i) = \sum_i \mu(a_i)$, (A.21)

for pairwise disjoint sets a_i, a_k, and

(c) $\mu(a \cap b) = 1$, if $\mu(a) = \mu(b) = 1$. (A.22)

The comparison with the corresponding properties (A.4) and (A.5) for the classical probability spaces shows that we need a third condition, (A.22), which in the classical case is a consequence of the others but not in the quantal case. Such probability measures are also called a "state" of the system, since the preparation of a state produces always a measure of this kind.

2.3. *The Random Variables*

Since we no longer dispose of the space Ω, only the second version of the notion of random variables can be transferred from the classical to the quantal calculus. Thus we define a random variable as a σ-homomorphism

$$\xi: \mathscr{B}(\mathbb{R}) \to \mathscr{L} \tag{A.23}$$

satisfying the three conditions

(a) $\xi(\emptyset) = \emptyset$, and $\xi(R) = I$, (A.24a)

(b) $\xi(\Delta_i) \perp \xi(\Delta_k)$, if $\Delta_i \perp \Delta_k$, for $i \neq k$, (A.24b)

(c) $\xi(\bigcup_i \Delta_i) = \bigcup_i \xi(\Delta_i)$. (A.24c)

An important consequence of this definition is that the range of any random variable ξ is a Boolean sub-lattice of $\mathscr{L}$, since the mapping ξ preserves the lattice structure, in particular the distributive law.

2.4. *The Distribution Functions*

The distribution function is defined, as in the classical case, by Equation (A.12).

2.5. *The Expectation Value*

In the definition of the expectation value also, we can take over the classical Equation (A.15).

From the foregoing we note the close analogy of the classical and quantal probability calculus. The difference is exhibited at only one point, the distributive law of the proposition system which is valid only in the classical case. It has important consequences.

In order to exhibit these consequences more explicitly, it is useful to introduce the notion of compatible propositions. Two propositions $a, b \in \mathscr{L}$ are said to be compatible if the lattice generated by a, b and its complements a', b' is Boolean. A sublattice $c\mathscr{L}$ is Boolean if and only if every pair of elements from $\mathscr{L}$ is compatible. Finally two random variables ξ and η are compatible if their ranges are. Thus if

$$\mathcal{L}_\xi = [x \mid x = \xi(\Delta)], \quad \text{and} \quad \mathcal{L}_\eta = [y \mid y = \eta(\Delta)] \quad \text{(A.25)}$$

then the two random variables ξ and η are compatible, if every $x \in \mathcal{L}_\xi$ is compatible with every $y \in \mathcal{L}_\eta$.

We see that the effect of the rejection of the distributive law is the new notion of *compatible random variables* which has no parallel in classical probability (or which is always trivially satisfied).

It is now easy to understand why operators in quantum

theory are represented by selfadjoint operators in Hilbert space. Indeed, in conventional quantum theory the lattice becomes the lattice of all projection operators in a Hilbert space (or equivalently the lattice of all closed subspaces). The map $\xi: \mathscr{B}(\mathbb{R}) \mapsto \mathscr{L}$ becomes what is known mathematically as a *spectral measure.* Every such measure determines uniquely a selfadjoint operator and vice versa. These operators are therefore concise representations of spectral measures, and hence of random variables in quantal probability calculus.

Another new feature shows itself when one considers joint distributions of two observables. If ξ and η are two observables (random variables) then

$$F_{\xi,\eta}(a, b) = \mu(\xi(S_a) \cap \eta(S_b)) \qquad \text{(A.26)}$$

is a non-decreasing function of both arguments. If ξ and η are compatible, $F_{\xi,\eta}$ satisfies

$$\int_{-\infty}^{+\infty} \mathrm{d}_b F_{\xi,\eta}(a, b) = F_\xi(a) , \qquad \text{(A.27a)}$$

and

$$\int_{-\infty}^{+\infty} \mathrm{d}_a F_{\xi,\eta}(a, b) = F_\eta(b) . \qquad \text{(A.27b)}$$

If ξ and η are not compatible then these last two conditions need no longer hold. In that case we may say that there does not exist a joint distribution for the two observables. This has been verified, for instance for the variables P and Q which satisfy the canonical commutation rules

$$[P, Q] = -i \,. \tag{A.28}$$

This phenomenon is an essential new feature of the quantal probability calculus.

REFERENCES AND NOTES

[1] W. Heisenberg, *Z. Phys.* **33**, 879 (1925); M. Born and P. Jordan, *Z. Phys.* **34**, 858 (1035); M. Born, W. Heisenberg and P. Jordan, *Z. Phys.* **35**, 557 (1926).

[2] P.A.M. Dirac, *Proc. Roy. Soc. (London)* **A 109**, 642 (1925); **A 110**, 561 (1926); **A 111**, 281 (1926); *Proc. Camb. Phil. Soc.* **23**, 412 (1926).

[3] E. Schrödinger, *Ann. Phys. (Leipzig)* **79**, 361, 489 (1926); **80**, 437, **81**, 109 (1926).

[4] L. de Broglie, *Ann. Phys. (Paris)* **3**, 22 (1925) and earlier papers, in *J. Phys. Rad.* and *Comptes Rendus* between 1922 and 1925.

[5] A. Einstein, *Sitz. Ber. Preuss. Akad. Wiss.* **1924**, pp. 261-267, **1925**, pp. 3-14, 18-25.

[6] W. Pauli, *Z. Phys.* **36**, 336 (1926).

[7] M. Born, *Z. Phys.* **37**, 863 (1926); **38**, 803 (1926); *Nachr. Gott. Akad. Wiss.* 1926, p. 146.

[8] P.A.M. Dirac, *Proc. Roy. Soc. (London)* **A113**, 621 (1926); P. Jordan, *Z. Phys.* **40**, 809 (1927).

[9] W. Heisenberg, *Z. Phys.* **43**, 172 (1927).

[10] N. Bohr, *Atti Congr. Int. Fisico Como*, 1927, Vol. II, pp. 565-588, reprinted in *Atomic Theory and the Description of Nature*, Cambridge (1934); *Nature* **121**, 580 (1928); *Naturwiss.* **16**, 245 (1928); Bohr's article on 'Discussion with Einstein on Epistemological Problems in Atomic Physics' in *Albert Einstein: Philosopher-Scientist* (ed. P.A. Schilpp), Library of Living Philosophers (1949).

[11] J. v. Neumann, *Mathematische Grundlagen der Quantenmechanik*, Berlin 1931 (English translation by R.T. Beyer, Princeton 1955).

[12] J.M. Jauch, *Helv. Phys. Acta* **37**, 293 (1964); J.M. Jauch and C. Piron, *Helv. Phys. Acta* **36**, 823 (1963); D. Finkelstein, J.M. Jauch, S. Schiminovich, and D. Speiser, *J. Math. Phys.* 3, 207 (1962); J.M. Jauch, *Helv. Phys. Acta* **33**, 711 (1960).

[13] P.A.M. Dirac, *Proc. Roy. Soc. (London)* **A114**, 243 (1927).

[14] P. Jordan and O. Klein, *Z. Phys.* **45**, 751 (1927); P. Jordan and E. Wigner, *Z. Phys.* **47**, 631 (1928).

[15] W. Heisenberg and W. Pauli, *Z. Phys.* **56**, 1 (1929); 59, 168 (1929/30).

[16] W. Heisenberg, *Z. Phys.* **77**, 1 (1932); D. Ivanenko, *Nature* **129**, 798 (1932); *Comp. Rend. (Paris)* **195**, 439 (1932); E. Fermi, *Z. Phys.* **88**, 161 (1934); H. Yukawa, *Proc. Phys. Math. Soc. (Japan)* **17**, 48 (1935); N. Bohr, *Nature* **137**, 344 (1936).
[17] N. Bohr, *J. Chem. Soc. (London)* **134**, 249 (1930) (Faraday Lecture); P. Jordan, *Anschauliche Quantentheorie*, Berlin 1936, Chapter 5; N. Bohr, *Atomic Physics and Human Knowledge*, New York 1958.
[18] W.P. Wigner, *Proc. Amer. Philos. Soc.* **113**, 95 (1969); *Foundation of Physics* **1**, 33 (1970).
[19] See 'Discussion with Einstein', quoted in Ref. 10.
[20] W. Heisenberg, *The Physical Principles of Quantum Theory*, Chicago 1930, (in German, Leipzig 1930); W. Heisenberg, *The Development of the Interpretation of the Quantum Theory*, in *Niels Bohr and the Development of Physics* (ed. W. Pauli), New York and London 1955. Heisenberg has always been opposed to a too narrow formulation and mathematical axiomatization of physical principles.
[21] W. Pauli, *Wellenmechanik*, in *Handbuch der Physik*, Vol. 33/I (ed. Geiger-Scheel), Berlin 1933, reprinted in *Handbuch der Physik*, V/1 (ed. S. Flügge), Berlin 1958. Two letters to Max Born, dated 3 March 1954 and 31 March 1954 respectively, in *The Born-Einstein Letters*, Walker and Company, New York 1971.
[22] G. Birkhoff and J. von Neumann, *Ann. Math.* **37**, 823 (1936); see also G. Birkhoff, 'Lattices in Applied Mathematics', *Proc. Symp. Pure Math.*, Vol. II Amer. Math. Soc. (1961), especially p. 155.
[23] L.H. Loomis, *Mem. Amer. Math. Soc.* **18**, 36 (1955); R. Scalettar, Thesis, Cornell University 1949 (unpublished); G. Ludwig, *Z. Phys.* **181**, 233 (1964).
[24] C. Piron, 'Axiomatique Quantique', *Helv. Phys. Acta* **37**, 439 (1964); cf. J.M. Jauch, *Foundations of Quantum Mechanics*, Addison-Wesley (1968), especially Section 8.3.
[25] J.M. Jauch, Varenna Lectures 1970.
[26] J.P. Eckmann and C.P. Zabey, *Helv. Phys. Acta* **42**, 420 (1969); S. Gudder and C. Piron, *J. Math. Phys.* **12**, 1583 (1971).

Strictly speaking, the representation theorem requires one more hypothesis, i.e. the irreducibility of the lattice: this means that one considers quantum systems not admitting superselection rules.

The representation theorem states that, under the conditions defined, there exists a field K (actually a division ring) with an involution $l \to l^*$ and a vector space V over K with an Hermitean (and Hilbertian form f such that the proposition lattice is isomorphic to the lattice of closed subspaces of V.

Now one knows from number theory that the only algebraic, complete fields are: (a) finite fields (or Galois fields); (b) p-adic fields (i.e. completions of the rationals with respect to the p-adic non-Archimedean valuation); (c) reals, complexes and quaternions (i.e. completions of the rationals with respect to the usual absolute value and their extensions; besides other algebraic extensions of them.

Gudder and Piron motivate the choice (c) by adding the assumption of the existence of "smooth observables". However, the same goal is simply achieved by ruling out the choices (a) and (b): this is done by Eckmann and Zabey (see Ref. 26) for the case (a), and by E.G. Beltrametti and G. Cassinelli (*Foundations of Physics* **2**, 1 (1972)) for the case (b). These two papers show that the involution $l \to l^*$ and the Hermitean form f do not exist if K is a finite field (this is intuitive) or a p-adic field (this is less intuitive). As for the complexes within the choice (c), we might refer to the papers by: E.C.G. Stueckelberg, *Helv. Phys. Acta* **33**, 727 (1960) and **34**, 621, 675 (1961) (reals are discussed), and D. Finkelstein and J.M. Jauch, *J. Math. Phys.* **3**, 207 (1963) (quaternions are discussed).

For a discussion of the representation theorem, also see Beltrametti and Cassinelli, 'On the Logic of Q.M.', Bonn University preprint (August 1972), to appear in a publication in honor of K. Bleuler.

27 P. Jordan, J. von Neumann and E. Wigner, *Ann. Math.* **35**, 29 (1934).

28 L. Rosenfeld, *Science Prog.* **163**, 393 (1953); L. Rosenfeld, 'Misunderstanding about the Foundation of Quantum Theory', in *Observation and Interpretation,* Proc. 9th Symp. of the Colston Research Society (Bristol 1957), (ed. S. Körner), London 1957 (reprinted by Dover, New York, 1962); L. Rosenfeld, *Nature* **181**, 658 (1958), **190**, 384 (1961); *Physics Today* **16**, no. 10, 45 (1963); *Nucl. Phys.* **A 108**, 241 (1968); *Progr. Theor. Phys. Suppl.*, p. 222 (1965).

29 In a series of papers Otto Halpern has studied the question whether one can actually reach the limit of uncertainty experimen-

tally. (O. Halpern, *Acta Phys. Austr.* **24**, 280 (1966); **28**, 353 (1968); **30**, 328 (1969); **33**, 305 (1971.)) He arrived at the conclusion that Heisenberg's discussion in his famous paper on the uncertainty principle (see Ref. 9) can be refined, and has obtained results which show that the uncertainty limit is higher than $h/4\pi$ which arises from the commutation relations. We do not share Halpern's opinion, as expressed in his last article (*Acta Phys. Austr.* **33**, 305 (1971), especially on p. 316), that his results have consequences for the interpretation of Q.M. The interpretation of Q.M. starts from the commutation brackets rather than from the specific numbers on the r.h.s. of the uncertainty relations.

[30] G.C. Wick, A.S. Wightman and E.P. Wigner, *Phys. Rev.* **88**, 101 (1952); *Phys. Rev.* D **1**, 3267 (1970). This point was recognized even earlier by Mme Destouches-Fevrier in her analysis of the logical structure of quantum mechanics, *La structure des Théories Physiques,* Presses Universitaires de France, Paris 1951.

[31] See also G.C. Hegerfeld, K. Kraus, and E.P. Wigner, *J. Math. Phys.* 9, 2029 (1968). In this paper the assumption of time reversal invariance on which the original proof was based has been dropped, and replaced by the assumption of rotational invariance.

[32] R. Mirman, *Phys. Rev.* **186**, 1380 (1969); Y. Aharonov and L. Süsskind, *Phys. Rev.* **155**, 1428 (1967).

[33] E.P. Wigner, *Z. Phys.* **133**, 101 (1952).

[34] H. Araki and M.M.,Yanase, *Phys. Rev.* **120**, 622 (1960).

[35] See E.P. Wigner, in 'Jubilee of Relativity Theory', (ed. A. Mercier/M. Kervaire), *Helv. Phys. Acta.,* Suppl. IV, Basel (1956), p. 210.

[36] H. Salecker and E.P. Wigner, *Phys. Rev.* **109**, 571 (1957).

[37] E. Schrödinger, *Sitz. Ber. Preuss. Akad. Wiss., Math. Phys. Kl.,* 238 (1931).

[38] M. Born, W. Heisenberg and P. Jordan, *Z. Phys.* **35**, 557 (1926).

[39] W. Heisenberg, *The Physical Principles of Quantum Theory,* Chicago 1930.

[40] W. Heisenberg and W. Pauli, Ref. 15.

[41] L. Landau and R. Peierls, *Z. Phys.* **69**, 56 (1931).

[42] N. Bohr and L. Rosenfeld, *Kgl. Danske Vid. Selks. Mat-Fys. Medd.* **12**, no. 8 (1933).

[43] N. Bohr and L. Rosenfeld, *Phys. Rev.* **78**, 794 (1950); E. Corinaldesi, *Nuovo Cim.* **8**, 491 (1951).

44 See E.P. Wigner, Observations at the end of the conference, *Contemporary Research in the Foundations and Philosophy of Quantum Theory* (ed. C.A. Hooker). D. Reidel, Dordrecht/Boston, 1973, pp. 369-385.
45 The statistical operator is a bilinear combination of wave functions.
46 See F. London and E. Bauer, 'La Theorie de l'Observation en Mecanique Quantique', *Actualites Scientifiques et Industrielles*, No. 775, Paris 1939.
47 See Ref. 44 and Section 4.4 of R.M.F. Houtappel, H. van Dam and E.P. Wigner, *Rev. Mod. Phys.* **37**, 595 (1965).
48 H. Margenau and R.N. Hill, *Progr. Theor. Phys.* **26**, 722 (1961).
49 E. Schrodinger, *Brit. J. Phil. Science* **3**, 109, 233 (1951).
50 P.K. Feyerabend, *Z. Phys.* **148**, 551 (1957). Also see his lecture 'On the Quantum Theory of Measurement', at the 1957 Bristol Symposium on *Observation and Interpretation in the Philosophy of Physics* (ed. S. Körner), London 1957, reprinted by Dover, New York 1962.
51 J.M. Jauch, *Helv. Phys. Acta* **37**, 293 (1964).
52 E.P. Wigner, *Am. J. Phys.* **31**, 6 (1962), B. D'Espagnat, *Suppl. Nueovo Cim.* **4**, 828 (1966).
53 A suggestion of E. Teller may also be interpreted in this direction. See E. Teller, 'Der quantenmechanische Messprozess und die Entropie' in *Werner Heisenberg und die Physik underer Zeit* (ed. F. Bopp), Braunschweig 1961.
54 W. Heisenberg, 'Die Zeit der Entwicklung der Quentenmechanik', in *Theoretical Physics in the Twentieth Century* (ed. M. Fierz and V.F. Weisskopf), Interscience Publishers, New York 1960.
55 Bohr stated in his Como lecture: "The very nature of the quantum theory thus forces us to regard space-time coordination and the claim of causality, the union of which characterizes the classical theories, as complementary but exclusive features of the description, symbolizing the idealization of observation and definition respectively." (see Ref. 10.)
56 W. Pauli, *Die allgemeinen Prinzipien der Wellenmechanik*, in *Handbuch der Physik*, Vol. 24/I, p. 89.
57 A quantitative evaluation of the degree of complementarity is possible, and was given by C. Piron (Ref. 24), and by J.M. Jauch and

G. Emch in 'Structures Logiques et Mathematiques en Physique Quantique', *Dialectica* **19**, 259 (1965). These authors show that for two elementary propositions a, $b \in L$ the quantity x, defined as

$$(a \cap b) \cup (a \cap b') \cup (a' \cap b) \cup (a' \cap b') = x\,,$$

is equal to I (the trivial proposition which is always true) if and only if a and b are compatible (or, if they are represented by projection operators they commute). If there exists a measure of the 'size' of an element x then this size would be a measure of the degree of incompatibility of the two propositions. Now such a measure always exists on the lattices and it is unique. It is the function ρ: $L \to R^+$ with the following properties

(1) $\rho(a) \geqslant 0\,,$

(2) $\Sigma\, \rho(a_i) = p(a_i)$ for $a_i \perp a_k\,,$

(3) $\rho(a \cup b) + \rho(a \cap b) = \rho(a) + \rho(b).$

We may then define $x \propto y$ if and only if $\rho(a) \leqslant \rho(b)$. The dimension functions of atomic lattices are positive integers. A pair of propositions a_1, b_1 is then less compatible than another pair a_2, b_2 if the corresponding x_1 and x_2 satisfy $x_1 \propto x_2$.

58 In his article on 'Questions of Method in the Consistency Problem of Quantum Mechanics' (*Nucl. Phys.* **A 108**, 242 (1968)), Rosenfeld stated: "It is clear that one cannot formalize the crucial point of the consistency argument. ... In particular, any axiomatization of the theory can at best help to avoid trivial contradictions or redundancies in its formal apparatus, but is incapable of throwing any light on the adequacy of the theory as a mode of description of experience. This last problem belongs to what the scholastics pointedly called 'real logic', i.e. logic of things, in contradiction to 'formal logic'."

59 Ref. 58, p. 243, Rosenfeld summarized: "The preceding indications, however fragmentary, should suffice to show that the difficulties under which the adepts of Von Neumann's method appear to labour are only consequences of this particular method, and do not occur if the problem is properly treated."

60 *The Born-Einstein Letters*, Walker and Company, New York,

1971; M. Born, *Kgl. Danske Vid. Selks., Mat.-Fys. Medd.* **30**, no. 2 (1955); *Phys. Bl.* **11**, 49 (1955); M. Born, *Bemerkungen zur statistischen Deutung der Quantenmechanik* (see Ref. 53), p. 103.
61 See in particular Born's paper in *Mat.-Fys. Medd.* (Ref. 60). Born studied thoroughly the example which he had obtained from Einstein. A modified example of the wheel has been discussed in the dialogue 'Are Quanta Real?' between Salviati, Sagredo and Simplicio, "recorded and translated into modern English" by J.M. Jauch, Indiana University Press, 1972.
62 F. Bopp, 'Statistische Mechanik bei Störung des Zustandes eines physikalischen Systems durch der Beobachtung', in: *Werner Heisenberg* (see Ref. 53).
63 D.I. Blokhintsev, *Quantum Mechanics*, Chapter 25, D. Reidel Publishing Co., Dordrecht, Holland, 1964.
64 Y. Aharonov, H. Pendleton and A. Petersen, *Int. J. Theor. Phys.* **2**, 213 (1969); Y. Aharonov and A. Petersen, 'Definability and Measurability in Quantum Theory'. in *Quantum Theory and Beyond* (ed. T. Bastin), Cambridge 1971. Also note Dirac's dictum: "Each photon interferes only with itself. Interference between two different photons never occurs"–P.A.M. Dirac, in *The Principles of Quantum Mechanics*, third edition, Oxford 1947, p. 9.
65 A. Einstein, *Verh. Deutsch. Phys. Ges.* **18**, 318 (1916); *Mitt. Phys. Ges. Zurich* **16**, 47 (1916).
66 N. Bohr, *Kgl. Danske Vid. Selsk., nat. og math. aft.* (8) 4.1., (1918).
67 E. Wigner, *Phys. Rev.* **40**, 749 (1932).
68 See Varenna Lectures 1970, lectures of J.M. Jauch and C. Piron. J.S. Bell has pointed out to me that he does not accept that the criticism about the physical meaning of $a \cap b$ (as set out by him before Bohm and Bub in his 1964 paper) was "effectively answered by Jauch and Piron".
69 J.E. Moyal, *Proc. Camb. Phil. Soc.* **45**, 99 (1948).
70 J.G. Gilson, *Proc. Camb. Phil. Soc.* **64**, 1061 (1968).
71 H. Margenau, in *Quantum Theory and Reality* (ed. M. Bunge), Springer, New York 1967.
72 E.P. Wigner, 'Quantum-Mechanical Distribution Functions Revisited', in *Perspectives in Quantum Theory* (Essays in honor of Alfred Landé, edited by W. Yourgrau and A. van der Merwe), M.I.T.

Press, Cambridge, Mass., 1971.
73 J.M. Jauch, private communication.
74 Details of the classical probability calculus are given in the Appendix. I am grateful to Professor Jauch for exciting and fruitful discussions on the notions of calculus in classical and quantum theories. Let us also mention H. Richter's *Wahrscheinlichkeitstheorie*, Springer, Berlin 1956.
75 Discussions with Professor Jauch. See Appendix.
76 P.A.M. Dirac, *Proc. Roy. Soc.* A 180, 1 (1942).
77 W. Pauli, *Rev. Mod. Phys.* **15**, 175 (1943).
78 S.N. Gupta, *Proc. Phys. Soc. (London)* A **63**, 681 (1950), and K. Bleuler, *Helv. Phys. Acta* **23**, 567 (1950) used the indefinite metric for longitudinal and scalar 'photons' in quantum electrodynamics. W. Heisenberg, *Z. Naturforsch.* **9a**, 292 (1954) and *Z. Phys.* **144**, 1 (1956) adopted it for the study of unified field theories. W. Pauli and G. Källen, (*Kgl. Danske Vid. Selsk. Mat.-Fys. Medd.* **30**, no. 7 (1955) found a state space with indefinite metric in the local form of a model of quantum field theory proposed by T.D. Lee, *Phys. Rev.* **95**, 1329 (1954).
79 W. Heisenberg first gave a consistent and physically interpretable version of the Lee model (*Nucl. Phys.* **4**, 532 (1957)). E.C.G. Sudarshan (*Phys. Rev.* **123**, 2183 (1961)) and E.C.G. Sudarshan and H. Schnitzer (*Phys. Rev.* **123**, 2193 (1961)) established a close connection between local Q.F.T. with indefinite metric and non-local Q.F.T. with positive metric. Applications of the indefinite metric have been made by Heisenberg and his collaborators (*Z. Naturforsch.* **14a**, 441 (1959)) to the unified field theory, by E.C.G. Sudarshan (*Nuovo Cim.* **21**, 7 (1961)) to weak interactions, and by M.E. Arons, M.Y. Han, and E.C.G. Sudarshan (*Phys. Rev.* **D2**, 1033 (1970)) to electrodynamics. The problem of the physical interpretation (and consistency) has been pursued vigorously during the past several years, and recent studies by E.C.G. Sudarshan and H. Rechenberg report encouraging results.
80 See C. Piron, quoted in Ref. 24, and J.M. Jauch, *Foundations of Quantum Mechanics*, also quoted in Ref. 24.
81 This is a consequence of Piron's representation theorem 24 (see Prion, Ref. 24). With respect to Q.F.T., however, it is not clear whether the assumptions and axioms of the Geneva group (Jauch and

collaborators) can be retained. In fact there are serious doubts whether they allow for a nontrivial relativistic interaction.

82 J.E. Moyal, quoted in Ref. 69.

83 E.P. Wigner, *Phys. Rev.* **40,** 749 (1932); also Ref. 72.

84 G. Ludwig, *Z. Phys.* **135,** 483 (1953).

85 H. Kummel, *Nuovo Cim.* **1,** 1057 (1955).

86 G. Ludwig, 'Gelöste und ungelöste Probleme des Messprozesses in der Quantenmechanik', in *Werner Heisenberg*, cited in Ref. 53.

87 A. Daneri, A. Loinger and G.M. Prosperi, *Nucl. Phys.* **33,** 297 (1962), *Nuovo Cim.* **44B,** 119 (1966). A similar idea was expressed, but not carried through, by E. Teller in his article cited in Ref. 53. for a more recent review of this work see G.M. Prosperi, 'Macroscopic Physics, Quantum Mechanics and Quantum Theory of Measurement in Quantum Theory and Beyond' (ed. T. Bastin), Cambridge 1971; see also the lectures at the 1970 Varenna Summer School.

88 A criticism of their work in this sense is also contained in the 'Comment on the Daneri-Loinger-Prosperi Quantum Theory of Measurement' by J. Bub, in *Quantum Theory and Beyond* (see Ref. 87).

89 Loinger does not assume the apparatus to be in a definite mixture at the beginning of the process (A. Loinger, *Nucl. Phys.* **A108,** 245 (1968). This seems to have been suggested in the criticism of J.M. Jauch, E.P. Wigner, and M.M. Yanase (*Nuovo Cim.* **48,** 144 (1967)).

90 P. Jasselette and J. Voisin have addressed themselves to this point in a recent article (*Rivista del Nuovo Cim.* **2,** 278 (1970).

91 L. Rosenfeld, in the paper cited in Ref. 58, has given it his full approval. Of course, it meets the general complementarity philosophy of Bohr but, as we have pointed out, it does not treat specific details and only implies a rather global statement in which many fine details may be observed.

92 E.P. Wigner has restated his discontent in the paper cited in Ref. 44: "The trouble is only that it postulates the miracle which disturbs us: that after the measurement, or if I use my own interpretation, after our observation, the apparatus is in a state which has a classical description. Hence, the explanation covers up rather than solves the problem. The transition to a classical description of the

apparatus is an arbitrary step which might obscure, but does not eliminate, the basic fact that the true equation of motion is deterministic." See also Jauch *et al.*, Ref. 89.

93 This particular form of the probability measure for a quantal state is a mathematical consequence of the σ-additivity of this measure on disjoint projections only. Thus, if E_i is a countable family of mutually disjoint projections ($E_i E_k = \delta_{ik} E_i$; or $E_i \perp E_k$ for $i \neq k$), then $\Sigma p(E_i) = p(E_i)'$ implies that $p(E)$ is of the form $p(E) = \mathrm{Tr}(WE)$. This is the theorem of A.M. Gleason (*J. Rat. Mech. Analysis* **6**, 885 (1957).

94 There is a review of the transition to macroscopic systems in the framework of C^*-algebra theory by K. Hepp, *Quantum Theory of Measurement and Macroscopic Observables,* dedicated to M. Fierz on the occasion of his 60th anniversary, Zurich preprint (see *Helv. Phys. Acta,* 1972).

95 C. George, I. Prigogine and L. Rosenfeld, *Kgl. Danske Vid. Selsk., Mat.-Fys. Medd.* **38**, no. 12 (1972).

96 The condition under which θ exists is very important, and invites careful study.

97 One should emphasize that the 'derivations' in this highly original program are often intuitive rather than mathematically rigorous. For instance, there are questions concerning the spectrum of H in relation to that of H_0, the meaning of 'correlations' etc. Prigogine and Grecos have given a rigorous formulation of the results recently, see *Proc. Nat. Acad. Sci.* **69**, 1629 (1972).

98 See A. Einstein, 'Bietet die Feldtheorie Möglichkeiten fur die Lösung des Quantenproblems?', *Sitz. Ber. Preuss. Akad. Wiss., Phys.-Math. Kl.* (1923), p. 359.

99 See L. Boltzmann, *Lectures on Gas Theory* (Leipzig 1896, 1898) translated by S. Brush, Berkeley 1964.

100 R. Fürth, *Z. Phys.* **81**, 143 (1933); D. Bohm and J.P. Vigier, *Phys. Rev.* **96**, 208 (1954). See also D. Bohm, 'A Proposed Explanation of Quantum Theory in Terms of Hidden Variables of a Sub-Quantum-Mechanical Level', in *Observation and Interpretation ...*, Ref. 50. Other hidden variable models have been discussed by L. de Broglie, *Comp. Rend.* **183**, 447 (1926), **184**, 273, **185**, 380 (1927); I. Fenyes, *Z. Phys.* **132**, 81 (1953), W. Weitzel, *Z. Phys.* **134**, 264, **135**, 270 (1933), **136**, 582 (1954).

101 E. Nelson, *Phys. Rev.* **150,** 1079 (1966); J.G. Gilson, *Proc. Camb, Phil. Soc.* **64,** 1061 (1968); J.E. Moyal, *Proc. Camb. Phil. Soc.* **45,** 99 (1949).
102 W.H. Furry, *Phys. Rev.* **40,** 393, 476 (1935). Furry considered only a very *particular* way of realizing Einstein's idea, and found it in disagreement with Q.M. (not just in the mathematical form, but as regards mathematical results. In 1965 J.S. Bell considered a much wider class, and reached the same conclusion. (See our Ref. 107).
103 A. Einstein, B. Podolsky and N. Rosen, *Phys. Rev.* **47,** 777 (1935).
104 D.Bohm, *Phys. Rev.* **85,** 166, 180 (1952); *Causality and Chance in Modern Physics,* Princeton 1957.
105 J.M. Jauch and C. Piron, *Helv. Phys. Acta* **36,** 827 (1963).
106 D. Bohm and J. Bub, *Rev. Mod. Phys.* **38,** 470 (1966); see also D. Bohm and J. Bub, *Rev. Mod. Phys.* **38,** 453 (1966).
107 J.S. Bell, *Rev. Mod. Phys.* **38,** 447 (1966); *Physics* **1,** 195 (1965).
108 J.F. Clauser, M.A. Home, A. Shimony, and R.A. Holt, *Phys. Rev. Lett.* **23,** 880 (1969).
109 S.J. Freedman and J.F. Clauser, *Phys. Rev. Lett.* **28,** 938 (1972); B. Misra, *Nuovo Cim.* **47,** 841 (1967).
110 S. Gudder, *Rev. Mod. Phys.* **40,** 229 (1968).
111 The experimental status of checking the prediction of Q.M. is not entirely conclusive, but no evidence has been obtained in favor or against Bell's theorem. The situation is summarized in the review article by V. Capasso, D. Fortunato, and F. Selleri, *Revista del Nuovo Cim.* **2,** 149 (1970). Note the results of the experiment of Freedman and Clauser (Ref. 108).
112 E.P. Wigner, *Amer. J. Phys.* **38,** 1005 (1970).
113 C. Papaliolios, *Phys. Rev. Lett.* **18,** 622 (1967).
114 *The Born-Einstein Letters* (Ref. 60). See letter dated May 2, 2952.
115 A. Einstein, *Dialectica* **2,** 320 (1948); Letters to Max Born, quoted in Ref. 114, Letters No. 88, 106 and 110.
116 E. Schrödinger, *Naturwiss.* **23,** 807, 823, 944 (1935). In this paper Schrödinger gave the famous example of the "cat paradox". The English version appeared in *Proc. Camb. Phil. Soc.* **31,** 555 (1935).

117 *Letters on Wave Mechanics* (ed. K. Przibram, translated by Martin J. Klein), Philosophical Library, New York 1967, p. 39.
118 N. Bohr, *Phys. Rev.* **48**, 696 (1935).
119 K. Hubner, 'The Philosophical Background of Hidden Variables in Quantum Mechanics', lectures at the Pennsylvania State University, September 1971. See also 'Ueber die Philosophie der Wirklichkeit in der Quantenmechanik' in *Philosophia Naturalis,* 1972.
120 Ref. 119, p. 5 of the English preprint.
121 Ref. 119, pp. 9-10 of the English preprint.
122 W. Heisenberg, Ref. 20, p. 65; also W. Heisenberg, *Daedalus* **87**, no. 3, 95 (1958).
123 E. Wigner, Ref. 44, p. 374.
124 E. Wigner, *The Monist* **48**, no. 2 (1964); reprinted in E.P. Wigner, *Symmetries and Reflections*, Indiana University Press, Bloomington, 1967, p. 185.
125 Ref. 124, pp. 190, 191.
126 Published in Moscow 1947 (in the series 'Questions of Philosophy', I).
127 G.F. Alexandrov, *Dokl. Akad. Nauk.* **84**, 1952.
128 V.I. Lenin, *Marxism and Empiro-Criticism,* Izdanie 'Zveno', Moscow, 1909.
129 The first edition of Blokhintsev's book (Ref. 62) appeared in 1944 in Moscow, when Stalin was at the height of his power. Howover one finds from recent writings of Blokhintsev that he has not altered his views too much.
130 D.I. Blokhintsev, Ref. 63, pp. 536-537. The reference to P. Jordan in this quotation is to his *Physics of the Twentieth Century,* Philosophical Library, New York 1944.

In another note of disapproval, Blokhintsev wrote: "According to Bohr, all problems of quantum theory have to be regarded as problems of the interaction between the apparatus and the micro-object, i.e. the interaction between subject and object – and that is where he abandons the firm foundation of physics. Herein lies also the fundamental methodological error of complementarity: in the light of this conception, the quantum mechanical phenomena lose their objective character, and begin to depend on the manner in which the human being perceives the phenomena of the micro-world. And this is *idealism.*" [D.I. Blokhintsev, 'Kritik der philosophischen

Anschauungen der sog. "Kopenhagener Schule" in der Physik' in: *Sowjetwissenschaft, Naturwissenschaftliche Abteilung* **6**, 4 (1953)]
131 V. Fock, 'Ueber die Deutung der Quantenmechanik', in Max-Planck-Festschrift 1958 (ed. by W. Kockel, W. Macke and A. Papapetrou), Berlin 1958, p. 117.
132 Ref. 151, p. 186.
133 *Are Quanta Real?*, Dialogue between Salviati, Sagredo and Simplicio, "recorded and translated into modern English" by J.M. Jauch, Indiana University Press, 1972.
134 I am grateful to Professor Jauch for discussing various aspects of this dialogue with me.
135 K. Baumann, *Z. Naturforsch.* **251**, 1954 (1970); see also H.D. Zeh, *Foundations of Physics* **1**, 69 (1970).
136 H. Everett, *Rev. Mod. Phys.* **29**, 454 (1957); J.A. Wheeler, *Rev. Mod. Phys.* **29**, 463 (1957); R.N. Graham, Ph.D. Thesis, University of North Carolina, 1971. See also B.S. Dewitt, *Physics Today* **23**, 30 (1970). The Everett theory is intelligible only in a form closely related to the de Broglie theory. (See J.S. Bell, 'On the "Impossibility" of Hidden Variables', Pennsylvania State University preprint 1971).
137 E.P. Wigner, 'Remarks on the Mind-Body Question', in *The Scientist Speculates* (ed. I.J. Good), London 1961, reprinted in *Symmetries and Reflections*, Bloomington 1967, p. 171. See also the article 'Epistemology of Quantum Mechanics', in *Contemporary Physics II* (ed. A. Salam), Vienna 1969, p. 431.
138 In his *Daedalus* article, cited in Ref. 122, pp. 99, 100, Heisenberg stated: "In consequence, we are finally led to believe that the laws of nature which we formulate mathematically in quantum theory deal no longer with the particles themselves but with our knowledge of the elementary particles ... Natural science always presupposes man, and we must become aware of the fact that, as Bohr has expressed it, we are not only spectators but also always participants on the stage of life."
139 See the papers cited in Refs. 44 and 47.
140 See Ref. 137, pp. 175-176.
141 Ref. 137, p. 176.
142 Ref. 137, p. 180.
143 See also Wigner's article 'Epistemology of Quantum Mechanics

– Its Appraisal and Demands', in *Psychological Issues VI,* No. 2, pp. 22-49. The difficulty in the usual description is clearly discussed there. It is the study of the measuring apparatus which leads to a chain or hierarhcy of new measuring devices. Wigner summarized: "This method of ascertaining the properties of the apparatus is, however, clearly outside the framework of physical theory and illustrates even more strikingly than the situation encountered with the reading of the apparatus that present physical theory is not self-contained ,that it constantly relies on everyday knowledge and is only an extension thereof." (p.34.)

Moldauer has recently reinterpreted Von Neumann's theory of measurement in quantum mechanics so that the experimental arrangement specifies the location of the "cut" by calling for the separate observation of the object and the measuring apparatus after the initial measurement interaction. The measurement ascertains which element of the mixture describing the final state of the apparatus is actually present. Moldauer has criticized the relevance and feasibility of observing the final coherent state of the object plus apparatus, and has discussed the "paradoxes" of "Schrodinger's cat" and "Wigner's friend" (P.A. Moldauer, *Foundations of Physics* **2,** No. 1, 41 (1972)).

144 See the article 'Epistemology of Quantum Mechanics', in *Contemporary Physics II,* cited in Ref. 137.

145 A. Einstein, *Phys. Z.* **10,** 185 (1909). See also Einstein's papers in *Verh. Deutsch. Phys. Ges.* **18,** 318 (1916) and *Phys. Z.* **18,** 121 (1917).

146 N. Bohr, quoted in Ref. 10.

147 See the discussions with O. Klein, Feb. 20, 1963, p. 9, *Archives for the History of Quantum Physics,* See also N. Bohr, Ref. 17, p. 96.

148 L. Rosenfeld, *Physics Today* **16,** 47 (October 1963).

149 N. Bohr, *Atomic Physics and Human Knowledge,* New York 1957, and N. Bohr, *Essays 1958-1962: on Atomic Physics and Human Knowledge,* New York 1963.

150 Address at the opening meeting of the International Congress on Light and Therapy in Copenhagen, August 1932, printed in *Nature* **131,** 421 (1933).

151 *l.c.,* Ref. 150, p. 11.

166 I. Prigogine, 1st International Conference on Theoretical Physics and Biology at Versailles 1967, (Amsterdam 1969), p. 23; I, Prigogine and P. Glansdorff, *Tehrmodynamics of Structure, Stability, and Fluctuations*, London 1971, Chapter XVII, I. Prigogine and R. Lefever, 'Thermodynamics, Structure and Dissipation', in *Biological Aspects of Electrochemistry*, see *Experientia Suppl.* **18**, 101-126, (1971); I. Prigogine, G. Nicolis, and A. Babloyantz, 'Thermodynamics of Evolution', *Physics Today*, November and December, 1972, pp. 23-28 and pp. 38-44 respectively; M. Eigen, *Naturwiss.* **58**, 465 (1971).
167 I am grateful to Professor Prigogine for illuminating discussions. For the most recent and complete statement of Prigogine's work, see his contributions to the volume *The Physicist's Conception of Nature*, ed. J. Mehra (D. Reidel, Dordrecht/Boston, 1973).
168 See Prigogine, Nicolis and Babloyantz, quoted in Ref. 166.
169 See Eigen, *l.c.*, p. 16.
170 E.P. Wigner, 'Are We Machines?', *Proc. Amer. Philos. Soc.* **113**, 95 (1969).
171 See Ref. 137, especially the discussion in Section I.
172 From *The Logic of Personal Knowledge*: Essays in Honor of Michael Polanyi (ed. E. Shils,) London 1961, p. 231, reprinted in 'Symmetries and Reflections', Ref. 137, p. 200. Our quotations are from the original article.
173 M. Eigen, quoted in Ref. 166.
174 M. Eigen, Ref. 166, p. 3.
175 P.T. Landsberg, *Nature* **203**, 928 (1964).
176 M. Ageno, *Nature* **205**, 1306 (1965).
177 E.P. Wigner and P.T. Landsberg, *Nature* **205**, 1307 (1965).
178 E.P. Wigner, *Foundations of Physics* **1**, 33 (1970).

GPSR Compliance

The European Union's (EU) General Product Safety Regulation (GPSR) is a set of rules that requires consumer products to be safe and our obligations to ensure this.

If you have any concerns about our products, you can contact us on ProductSafety@springernature.com

In case Publisher is established outside the EU, the EU authorized representative is:

Springer Nature Customer Service Center GmbH
Europaplatz 3
69115 Heidelberg, Germany

Batch number: 10830935

Printed by Printforce, the Netherlands

THE QUANTUM PRINCIPLE:

ITS INTERPRETATION AND EPISTEMOLOGY